Greenhouse Production

RONALD J. BIONDO

Agricultural Educator

Naperville, Illinois

GREENHOUSE PRODUCTION

AgriScience & Technology Series

Jasper S. Lee — Series Editor

PEARSON
Prentice Hall Interstate

Upper Saddle River, New Jersey

ISBN 0-13-036422-3

2 3 4 5 6 7 8 9 10 07 06 05 04

Preface

Greenhouse Production is a comprehensive book based on the greenhouse industry. It takes the reader beyond the most basic levels of horticulture. It includes botany and the production of floriculture crops. Greenhouse structures, equipment, and management techniques are covered, and specific growing instructions for the different floriculture crops are provided. Employment skills and business aspects of a greenhouse operation are also included.

This book represents a new approach in horticulture education. It is intended to be exactly what is needed in today's horticulture programs for greenhouse production. A student- and teacher-friendly approach is used. An abundance of color photographs and line drawings enrich the book by illustrating concepts discussed in the text. Activities are provided that go beyond the classroom.

Acknowledgments

I would like to acknowledge numerous people for their contributions to this book. Special recognition goes to Dr. Marvin Carbonneau, Professor Emeritus, University of Illinois, for his instruction and to Mike Schaefer, of Schaefer Greenhouses, Inc., for his open-door policy and the time spent discussing greenhouse crop production.

Many others deserve thanks for their contributions, including

- Tom Cosgrove, *Greenhouse Product News*
- Lim-King Brady, Crystal Company
- Louis D'Amours, Berger Peat Moss
- David H. Johnson, United Greenhouse Systems
- Roy Klehm, Klehm Nursery
- Jennifer Sparks, Society of American Florists
- Brian Corr and Ginny Kovach, Ball Horticultural Company
- Arthur Clesen, Inc.
- Steven E. Newman, Extension Greenhouse Specialist, Colorado State University
- Dave Thill, Mid-American Growers
- Andre Rutte, DeVroomen Holland Garden Products
- Kuhn Flowers
- Carl Reed, Barrington High School (Illinois)
- Alice White-Klinkhamer, International Flower Bulb Centre
- Christine Kelleher, Yoder Brothers, Inc.

- CO-EX Corporation
- Tim Landers, Benchmark Foliage
- Randy Strode, Agristarts, Inc.
- Mike Soucek, Natural Beauty of Florida
- Marty Knox, Knox Nursery, Inc.
- Greg Stack, Fredric Miller, and James E. Schuster, University of Illinois Extension
- Joe Steffen, Newark High School (Illinois)
- Jim Frost, Urban Floral, Inc.
- Ludvig Svensson, Inc.
- Hank Maday, Maday Greenhouses
- Mike Cherim, The Green Spot, Ltd.
- The Geiger Companies
- USDA Agricultural Research Service
- Gary Mangum, Bell Nursery
- Ken Butler, Scotland
- Red Kennicott, Kennicott Florists
- Dianne Noland, University of Illinois
- National FFA Organization
- California Cut Flower Commission
- Greg Goff, Ball Seed Company
- Vernie Thomas, Vero-Media, Inc.

Special appreciation is due to Ester V. Biondo for her sharp proofreading and to my wife, Wendy, and my sons, Steven and Thomas, for their patience and support. Of course, much credit goes to Dan Pentony, Ron McDaniel, Kim Romine, and Rita Lange, of the Center for Agricultural and Environmental Research and Training (CAERT), Inc., for their support and for the incredible amount of work required to put the final product in print.

Ronald J. Biondo

Contents

GREENHOUSE PRODUCTION

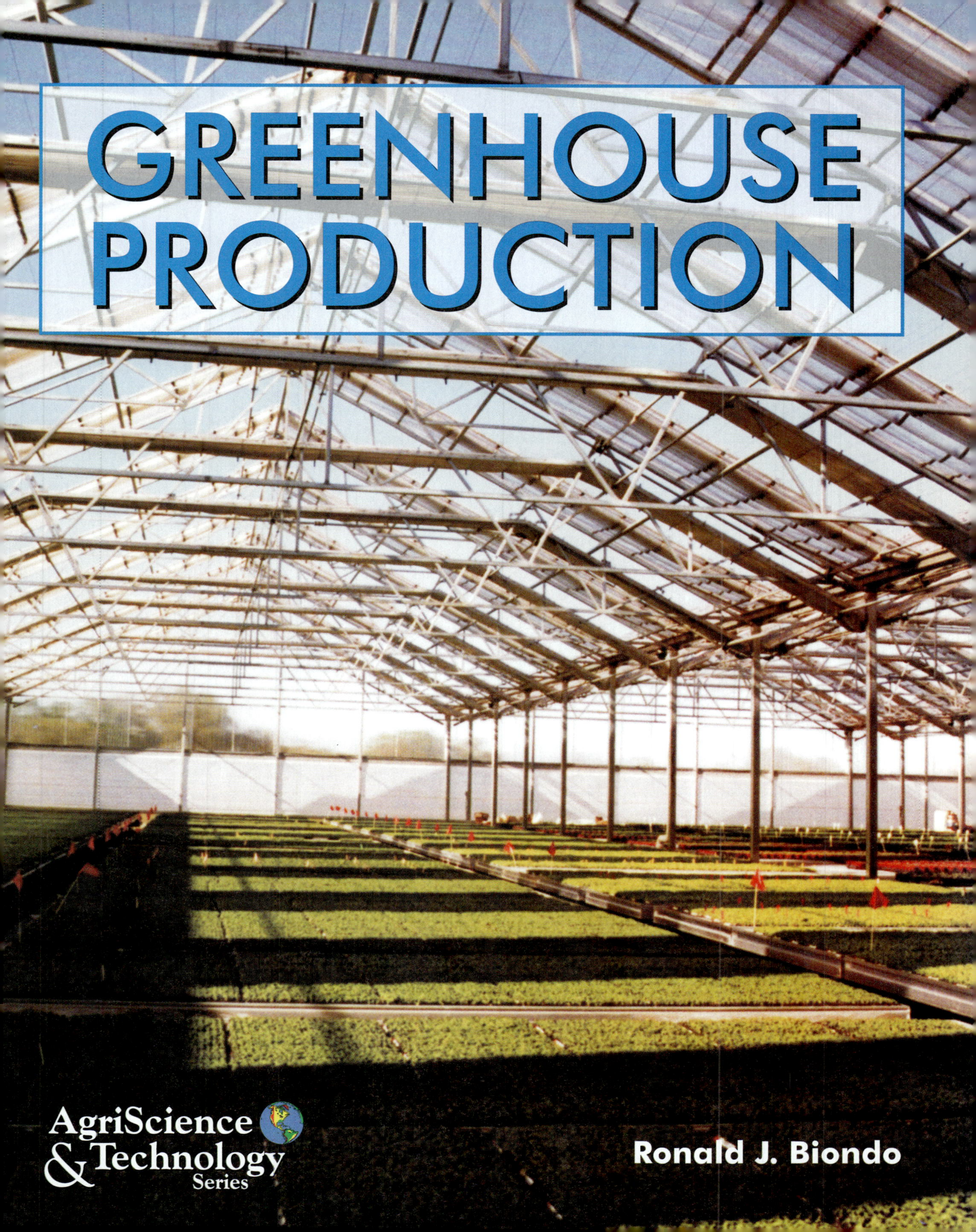

AgriScience & Technology Series

Ronald J. Biondo

CHAPTER 1

The Floriculture Industry

OBJECTIVES

This chapter is an overview of the floriculture industry. It has the following objectives:

1. Describe the scope of the floriculture industry.
2. Explain the importance of the greenhouse industry.
3. Describe the extent and value of the floral design industry.
4. Identify careers found in the floriculture industry.

TERMS

bedding plants
buyer
conditioning
cut flowers
cut cultivated greens
floral arranger
floral designer
floral production
floriculture
foliage plants
greenhouse
greenhouse worker
grower
marketing manager
ornamental horticulture
plant propagation
potted flowering plants
propagator
retail floral manager
retail florist
salesperson
wholesale floral manager
wholesale florist
vase life

1–1. There are many floral products. (Courtesy, Society of American Florists)

QUICK! Mother's Day is approaching and you need to show your mother that you care. You could buy some candy, but somehow candy doesn't express your feelings. You could buy a card, but that doesn't appear to be enough. If you are like many people, a floral product seems to be most appropriate. Now, you must choose a bouquet, a houseplant, a corsage, a potted azalea, or one of many other floral products that are available. In fact, the number of items from which to choose can make the decision difficult. The good news is you have a wide selection of choices.

Have you ever given thought to where all these floral products come from? Have you considered how these products are grown so they are available at specific times of the year? Have you ever marveled at the ability some people have in designing floral arrangements? Most people don't give these things much thought. They simply appreciate the availability of the floral products. However, people who pose these questions touch upon the exciting floriculture industry.

WHAT IS THE SCOPE OF THE FLORICULTURE INDUSTRY?

The literal definition of ***floriculture*** is the "culture of flowers." Floriculture is based on flowering and foliage (leafy) plants. It is an international, multibillion dollar industry. Floriculture businesses include the production of floral crops, the distribution of the crops from the grower to the consumer, and the processing of crops before sale. Some common floriculture businesses are retail and wholesale production greenhouses, florist shops, and floral supply companies. Floriculture is a major part of a larger industry called ornamental horticulture. ***Ornamental horticulture*** is the practice of growing and using plants for decorative purposes. Another major part of ornamental horticulture includes the nursery/landscape industry.

1–2. Floriculture is the "culture of flowers."

FLORAL PRODUCTION

Floral production involves the growing of flowering or foliage crops to sale size or maturity. The majority of the crops are produced in greenhouse structures. The value of floriculture crop production in the United States has shown steady growth. The United States Department of Agriculture (USDA) estimates the total wholesale crop value for all growers with $10,000 or more in sales at $4.74 billion for 2001. The states of California (22 percent) and Florida (16 percent) lead the way in floriculture production. Amazingly, the top five states—California, Florida, Texas, Michigan, and Ohio—accounted for $2.52 billion, 53 percent of the total value.

Table 1–1. Floriculture Production in the United States Based on Operations in Excess of $100,000 in Wholesale Sales. (USDA Floriculture Crops 2002 Summary)

State	Percentage of Sales
California	23%
Florida	17%
Texas	6%
Michigan	6%
Ohio	4%

When greenhouse operations having sales in excess of $100,000 are considered, some interesting facts surface. These larger greenhouse operations dominate the market. Of all the greenhouse operations surveyed by the USDA, only 43 percent had sales above $100,000. Yet, they accounted for $4.44 billion in 2001 or 94 percent of the total value of floriculture crops.

What are the floriculture crops being grown? The wholesale value of bedding and garden plants reached $2.18 billion in 2001. That represents 46 percent of the wholesale value of all floriculture crops. Potted flowering plants were valued $832 million. The foliage category accounted for $585 million. The value of cut flowers was $424 million, while cut cultivated greens were valued at $111 million.

Table 1–2. The Wholesale Value of Top Floriculture Crops in the United States in 2001 (USDA Floriculture Crops 2002 Summary)

Crop	Sales	Percentage
Bedding and garden plants	$2.18 billion	46%
Potted flowering plants	$832 million	18.7%
Foliage plants	$585 million	13.2%
Cut flowers	$424 million	9.5%
Propagative material	$306 million	6.9%
Cut cultivated greens	$111 million	2.5%

CAREERS IN THE GREENHOUSE INDUSTRY

Floriculture production involves people with different skills. Some have a sound understanding of plants and plant growth. Others may be strong in business. There is also room for

1–3. Growers are in charge of crop production. (Courtesy, *Greenhouse Product News*)

unskilled laborers. Some job titles associated with floriculture production include grower, propagator, marketing manager, and greenhouse worker.

Grower

The person in charge of crop production is called a ***grower***. Large greenhouse operations will employ a number of growers, each of whom may focus on one or more specific crops. Growers have extensive knowledge in crop production and are responsible for crops. Growers assign some tasks to greenhouse workers.

Propagator

Production begins with some form of plant reproduction, whether by seed or by using a portion of a plant. Plant reproduction is commonly known as ***plant propagation***. The new or young plants are grown to a size or an age at which they may be sold for further production purposes. Some might be sold directly to the consumers. A ***propagator*** is an employee trained in plant reproduction.

Greenhouse Worker

Greenhouse workers often start as unskilled laborers and are trained in greenhouse production tasks. Greenhouse crop production is labor intensive and there are plenty of opportunities for workers. The jobs assigned to a greenhouse worker might range widely. Some jobs include potting, disbudding, watering, applying pesticides, packaging crops for shipment, and maintaining greenhouse structures and equipment.

Marketing Manager

The ***marketing manager***, sometimes called a wholesale manager, has the responsibility of moving the crop to the customer. He or she promotes the sales of crops and handles orders. The marketing manager sees that crops are graded and selected for the customer, stored if necessary, packaged, and delivered.

GREENHOUSES

Most floriculture crops are produced in some type of greenhouse structure. A ***greenhouse*** is a structure enclosed by glass or plastic that allows light transmission for plant growth. Greenhouses give growers the ability to control environmental conditions affecting a crop. Greenhouses allow a high level of light to reach the plants. They can be heated or cooled depending on the needs of the plants. Watering and fertilizer applications are also carefully controlled by the grower. In addition to providing control of specific growing conditions, greenhouses allow plants to be grown throughout the year.

Greenhouse growers often specialize in growing certain crops. The decision to grow certain crops is based on a number of factors. The most important factor determining which crops are produced is profit. Can the crop be sold for a profit? The climate is another very important factor. The greatest concern with climate is the amount of sunlight and the average temperature. One reason the major producers of cut carnations and roses are located in Columbia, South America, is because the climate is perfect for growing these crops and greenhouses are unnecessary.

Nearness to the market influences decisions as to what crops are grown. Cut flowers can be grown far from the marketplace. They are relatively lightweight. As a result, the cost of

1–4. Greenhouse structures allow light transmission for plant growth. Pictured is the conservatory at Kew Gardens, England.

shipping cut flowers by air, even across oceans, is affordable. Potted plants are most often grown locally because of the cost of shipping. Laws also restrict the movement of soil across borders.

MAJOR TYPES OF FLORICULTURE CROPS

The floriculture crops produced in the greenhouse industry fall under a number of groupings. The largest of these groupings include cut flowers, cut cultivated greens, potted flowering plants, foliage plants, and bedding and garden plants.

1–5. Florists work with cut flowers and foliage. (Courtesy, Society of American Florists)

1–6. The poinsettia is a popular potted flowering plant. (Courtesy, Paul Ecke Poinsettias)

Cut flowers—Some floral production operations focus on supplying cut flowers to wholesale florists. That is, they grow flowers, cut them when they reach a certain maturity, and sell them to a wholesaler. Roses, carnations, chrysanthemums, and orchids are a few common cut flowers.

Cut cultivated greens or leaves grown for floral design work are also grown and harvested. Leatherleaf, lemonleaf, and huckleberry are common foliage materials used in floral work. Those in the industry often refer to cut cultivated greens as cut foliage or "greens."

Potted flowering plants—A large segment of the floriculture industry is the production of flowering plants in pots. The plants may be propagated by seed or through asexual means, such as cuttings of plants. The young plants are grown to the flowering stage in a greenhouse structure. The entire plant and the pot are then shipped to market. Some popular potted flowering plants are poinsettias, chrysanthemums, Easter lilies, and African violets.

1–7. Florida is the largest producer of foliage plants. (Courtesy, *Greenhouse Product News*)

Foliage plants—Some growers produce potted plants to be sold as foliage plants. Foliage plants are grown for their leaves rather than for their flowers. They are also called houseplants. Florida is a large producer of foliage plants, which are shipped throughout the United States. Philodendrons, dieffenbachias, figs, scheffleras, and dracaenas are common foliage plants produced for use in homes and offices.

Bedding and garden plants—***Bedding plants*** are herbaceous, annual flowers and vegetables. Also included in this category are herbaceous perennials and flowering hanging baskets. Most bedding and garden plants are produced in the spring for outdoor planting.

Many bedding plants are started by seed and grown in a greenhouse during the late winter and spring. They lack the ability to survive freezing temperatures and, therefore, must be planted outdoors after the risk of spring frost. Gardeners transplant the fairly mature plants in the landscape. By producing the plants well before a safe planting date, gardeners are given a longer period during which they can enjoy the plants. Bedding plants include impatiens, petunias, marigolds, tomatoes, and many other plants.

1–8. This person checks bedding plant orders for a large commercial greenhouse operation.

1–9. Bedding plants outsell all other floriculture crops.

Herbaceous perennials and flowering hanging baskets may or may not be propagated from seed. Examples of herbaceous perennials include daylilies, hardy garden mums, and hostas. Popular flowering hanging baskets are ivy geraniums, impatiens, and petunias.

FLORAL DISTRIBUTION

There are several channels in which floral crops can be distributed. The shortest route is from the local grower to the local consumer. The consumer may make a purchase at the production site or the grower may transport the plants to a nearby retail outlet. The weight of potted plants makes shipping expensive. Therefore, a majority of potted plants are produced close to the market. Restrictions in shipping soil across state and national borders is another reason potted plants are grown close to the market.

The longest distribution channel involves the international market. Although the distance involved in the international market is great, the distance itself seems less important than the time it takes to get the products to buyers. It is critical that floral products are delivered to market as soon as possible. Regular air traffic linking the world today makes the international market possible. In the international market, crops move from the producer to an exporter, to a wholesaler, to a retailer, and finally to the consumer. The scope of the international floral market is truly remarkable.

At one time, nearly all cut flowers sold in the United States were grown in the United States. Today, imports account for most of the major cut flowers sold in the United States. Ninety percent of the carnations, 71 percent of the roses, and 89 percent of the cut chrysanthemums sold in the United States are produced in other countries.

The center of the international wholesale market for cut flowers and foliage is located in Holland. In fact, Holland has long been recognized as the leader of the floriculture industry. Other countries that are major exporters or importers of floral crops include Columbia, Israel, Italy, Kenya, and the United States. Carnations grown in Columbia, roses grown in Kenya, and exotic flowers grown in Hawaii are flown to the flower auctions in Holland. Buyers from around the world inspect and buy the flowers at the auctions. Once purchased, the flowers are shipped by air or road to wholesale businesses. The wholesaler repackages the flowers and takes orders from retail florists. Because flowers are perishable, they must be bought and shipped very quickly.

Not all flowers go through Holland. A large number are shipped directly to the United States. In 1999, Columbia and Ecuador shipped a combined 134,000 tons of flowers worth $470 million to Miami. The following example traces fresh cut roses from Columbia to a retail floral shop in the United States.

1–10. Exotic flowers are distributed around the world. (Courtesy, Agricultural Research Service, USDA)

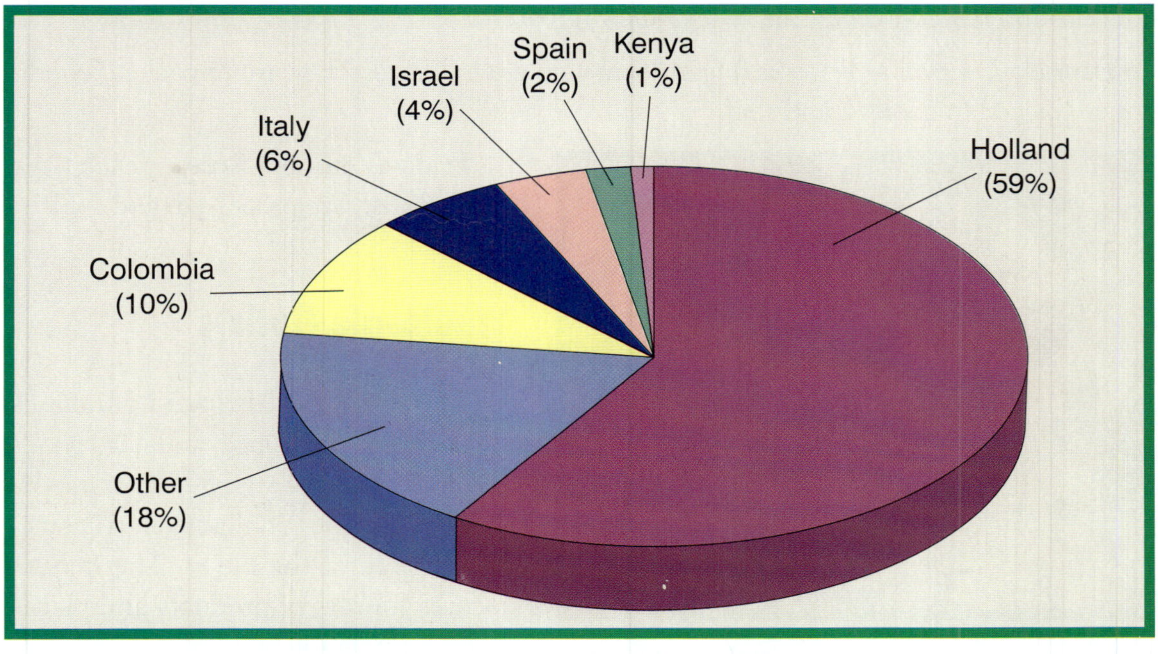

1–11. Percentage of world flower exports by country.

1–12. Holland is recognized as the leader in the floriculture industry. Shown is an international flower auction in Aalsmeer. (Courtesy, Floyd Giles)

Tuesday, 7 a.m.—Roses are cut early in the morning while well-hydrated and moved quickly to coolers.

Tuesday, 1 p.m.—Roses are graded by color, stem length, and stem thickness. They are bathed in a weak bleach solution to retard bacterial and fungal growth.

Wednesday, 6 a.m.—Roses are boxed, sometimes with ice, and trucked to the airport in Bogotá.

Wednesday, 9 a.m.—Roses arrive in Miami. Floral wholesalers ship the roses to destinations around the country by air, truck, and railroad.

Thursday, 4 p.m.—Roses arrive at local wholesale florists. They are unpacked, conditioned, placed in coolers, and sold.

1–13. Wholesale florists buy and sell flowers. (Courtesy, Society of American Florists)

Friday, noon—Roses reach the retail florist, are processed, and sold.

Wholesale Florists

The ***wholesale florist*** purchases cut flowers, cut foliage, and hard goods from a number of suppliers. An important job in the wholesale florist business is that of the buyer. The ***buyer*** locates sources of products and places bids or offers for purchase. The buyer must be very knowledgeable because the variety of available cut flowers and foliage numbers in the thousands. Also, the floral materials are often only available during certain seasons of the year. The hard goods purchased include ribbon, vases, hand tools, tape, and numerous other products used by florists. The wholesale florists resell and distribute the goods to retail businesses or to other wholesalers.

A ***wholesale floral manager*** oversees the operations of the business. He or she supervises the staff, estimates retail demands for products, and prepares long-range plans. Another position of importance is the salesperson. The ***salesperson*** has direct and usually frequent contact with the retail florist. The salesperson takes orders and arranges delivery of the products. Wholesale florists generally do not sell directly to the public.

1–14. The salesperson takes orders and arranges delivery of the products.

1–15. These workers in a wholesale florist package orders for shipping to retailers.

Retail Florists

Floriculture crops are sold to consumers through retail outlets. Traditionally, the ***retail florist*** provides floral design services, cut flowers, foliage plants, and potted flowering plants. Retail florists have typically been small independent businesses. Along with the retail florists, greenhouses and garden centers are recognized sources of cut flowers, foliage, and potted flowering plants. However, sales have expanded greatly to other retail outlets including supermarkets, gas stations, and large retail stores, such as Wal-Mart, Home Depot, and Kmart. These new outlets have made the purchase of floriculture crops more convenient for many people.

The ***retail floral manager*** has the responsibility of coordinating the operations of the business. Tasks associated with this position include managing personnel, working with budgets, and maintaining the store's inventory.

1–16. Retail florists display their products in attractive ways.

Highly skilled positions in the florist trade involve floral arrangers and designers. These people are artists that use flowers and other materials in their artwork. ***Floral arrangers*** have the training required to make and copy floral designs. ***Floral designers*** are considered to be more skilled than arrangers in that they have the ability to create and design unique floral art.

Given a customer order, arrangers and designers choose flowers and the necessary floral supplies to construct floral arrangements. Floral work may involve fresh cut flowers, cut cultivated greens, everlasting flowers (both dried and silk), and other materials. Flowers are selected and used to express feelings of love and sorrow, as well as to decorate homes and offices. Experienced florists offer helpful advice to customers on flower selection, care, colors, and costs. Special occasions, such as anniversaries, weddings, parties, holidays, and funerals, make up the bulk of floral sales.

1–17. Floral designers create designs. (Courtesy, Kuhn Flowers, Jacksonville, FL)

1–18. Flowers are given to celebrate special occasions, such as the birth of a child, or two.

FLORAL PROCESSING

Floral processing involves the handling and preparation of floral products for sale. Processing occurs primarily with the cut flower industry. Growers process floral crops for shipping. Usually, the wholesaler conditions cut flowers and cut cultivated greens upon receipt of the materials. **Conditioning** involves treatments that extend the life of the flowers. Further processing of cut flowers and cut cultivated greens is done by retail florists before sale to the public.

It is important for cut flowers to be moved from the grower to the consumer as quickly as possible. That is because flowers begin to die once they are cut from the plant. Handling the flowers involves keeping them cool. Lower temperatures slow the life processes of the flowers. Flowers are also placed in clean water with a floral preservative. The preservative provides the flower with food and reduces the growth of bacteria that can clog conductive tissues. The proper handling of flowers serves to extend their vase life. **Vase life** is the length of time cut flowers and cut cultivated greens live after they have been cut.

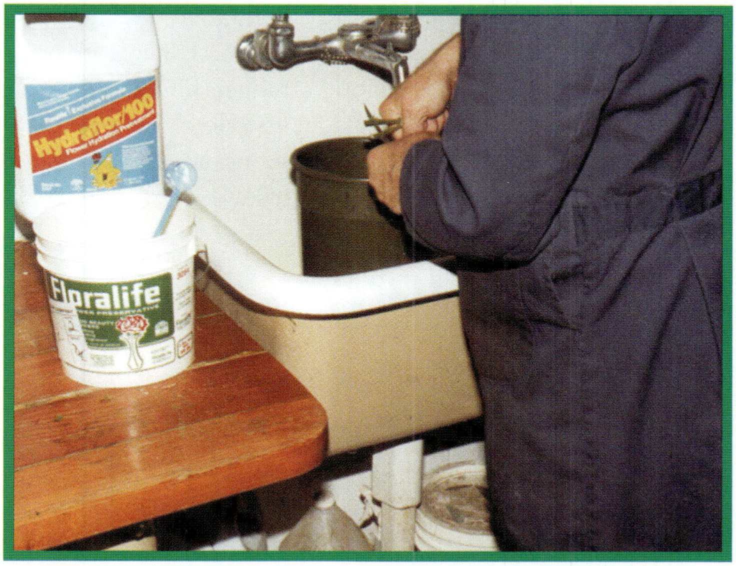

1–19. Flowers are conditioned to extend their lives.

REVIEWING

MAIN IDEAS

Floriculture is part of the ornamental horticulture industry. It includes those areas pertaining to the production and use of flowering and foliage (leafy) plants. Major business components of the floriculture industry are the production of floral crops, the distribution of crops from the grower to the consumer, and the processing of crops before sale.

Floral production involves the growing of flowering or foliage crops for sale. The value of floriculture crop production in the United States has shown steady growth. California accounts for 22 percent and Florida accounts for 16 percent of floriculture production in the United States. Production begins with plant propagation. Growers grow the plants and most often use greenhouse structures. Greenhouses allow the control of environmental factors. Profitability, climate, and distance to market influence which crops are grown in a particular operation. The major types of plants produced are cut flowers, cut cultivated greens, potted flowering plants, foliage plants, and bedding plants.

Distribution of floriculture products is an important part of the industry. Consumers often purchase potted flowering plants, foliage plants, and bedding plants at the production site or at a retail outlet near the production site. The international market is important in terms of cut flowers and cut cultivated greens. In the international market, crops move from the producer to an exporter, to a wholesaler, to a retailer, and finally to the consumer. Today, imports account for 90 percent of the carnations, 71 percent of the roses, and 89 percent of the cut chrysanthemums sold in the United States.

Floral processing involves the handling and preparation of floral products for sale. Proper handling of flowers serves to extend their vase life or the length of time cut flowers and foliage live after they have been cut. Cut flowers and foliage are processed by florists—people who sell cut flowers and floriculture crops. Floral arrangers and designers choose flowers and the necessary floral supplies to construct floral arrangements. Floral arrangers have the training required to make and copy floral designs. Floral designers are considered to be more skilled than arrangers in that they have the ability to create and design unique floral art.

QUESTIONS

Answer the following questions using complete sentences and correct spelling.

1. What is floriculture?
2. What is floral production?
3. What is the role of a grower?
4. Why are greenhouses beneficial in producing crops?
5. What factors influence the decision to produce certain crops?
6. What are the major floriculture crops? Define each.
7. How has air travel changed the floriculture industry?
8. What is vase life?
9. How do floral arrangers and floral designers differ?
10. What skills do florists call upon besides making arrangements?

EVALUATING

Match the term with the correct definition. Write the letter of the term on the line provided.

a. greenhouse	e. plant propagation	i. grower
b. bedding and garden plants	f. foliage plant	j. ornamental horticulture
c. floral designer	g. floriculture	k. floral production
d. potted flowering plants	h. retail florist	l. vase life

______1. Considered to be more skilled than a floral arranger in that he or she has the ability to create and design unique floral art.

______2. The growing of flowering or foliage crops to sale size or maturity.

______3. A potted plant, sometimes called a houseplant, grown for its leaves rather than for its flowers.

______4. The length of time cut flowers and cut cultivated greens live after they have been cut.

______5. Poinsettia, chrysanthemum, Easter lily, and African violet plants grown, shipped to market, and sold in a pot.

______6. The "culture of flowers."

______7. A business that provides floral design services, cut flowers, and other floriculture crops.

______8. The practice of growing and using plants for decorative purposes.

______9. Plant reproduction.

_____10. A structure enclosed by glass or plastic, which allows light transmission for plant growth.

_____11. The person in charge of the crop production.

_____12. Herbaceous, annual flowers and vegetables for outdoor planting.

EXPLORING

1. Make an appointment with a local greenhouse grower or florist for a 10 minute interview. Prior to the interview, compose a list of questions that interest you about their work. Dress appropriately and conduct the interview in a professional manner. Who knows, your effort may open doors for future employment.

2. Visit a retail floral outlet. Record all the different types of plants for sale. Investigate further to determine where they were produced. Look into how they were produced and how they ended up at the retail outlet.

3. Ask your school administration and instructor to allow you to job shadow for a day. Then, identify a wholesale florist, a greenhouse grower, a floral designer, or similar person who you can observe for a day.

CHAPTER 2

Employability in the Greenhouse Industry

OBJECTIVES

This chapter has the following objectives:

1. Report on desirable workplace skills and expectations.
2. Discuss general safety rules in the workplace.
3. Recognize the value of supervised experience programs.
4. Explain related opportunities found in student youth organizations.

TERMS

attitude
career success
cooperation
dependability
enthusiasm
honesty
initiative
leadership
loyalty
motivated
National FFA Organization
National Junior Horticultural Association (NJHA)
personal growth
Supervised Agricultural Experience (SAE)

2–1. These people enjoy their work. (Courtesy, The National FFA Organization)

WORK. What is your view of work? Have you ever known someone that truly enjoys work? Have you ever seen someone that dreads work? Everyone knows people that enjoy their work and others that seem to be depressed by work. Not surprisingly, the ones who appear to enjoy work are often successful. They are praised and rewarded financially.

Success in work often depends on a person's outlook. There are things you can do to increase your chances of being successful. Whether you work for a large company or a small mom and pop outfit, the issues remain the same. This chapter focuses on things expected of the employee, expectations of the employer, and some ways young people can prepare themselves for a positive career.

SUCCESS IN THE WORKPLACE

Employment comes with expectations. What does the employer expect of employees? What do employees expect from the employer? What do employees expect from each other? Understanding these questions and recognizing answers to them can lead to successful and satisfying employment.

2–2. A person with a positive attitude smiles easily and is cheerful. (Courtesy, The National FFA Organization)

ATTITUDE

Attitude can be defined as one's outlook on life. It is how one views the world around them. Attitude can be the greatest factor in job success. A person is generally considered to have one of two types of attitude. He/she has a positive attitude or a negative one.

People with positive attitudes smile easily and are cheerful. They are outgoing and tend to see life as exciting and enjoyable. In the workplace they rarely complain, seldom criticize others, respect the opinions of others, and accept responsibility for mistakes. People with positive attitudes make good team players.

Those people with a negative attitude appear unhappy. They tend to complain a lot, find fault in others, make excuses for mistakes, are self-centered, and rarely smile. Often people with bad attitudes find it difficult to work cooperatively with others.

Employers and employees are subject to positive and negative attitudes. Employers like employees that display a positive attitude. They are easy to work with and promote a positive image of the business. By the same token, employers with positive attitudes are a pleasure to work for and provide a positive work atmosphere.

EMPLOYER'S EXPECTATIONS

It is interesting to hear from employers as to what they expect from the people who work for them. The readiness of young people for work often surfaces as the top concern expressed by many employers. Readiness for work centers around workplace skills, those things expected of an employee. Employers seldom are concerned about the person's interest in growing plants or the person's greenhouse skills. Most employers believe they can teach an employee the necessary greenhouse technical skills.

Employer's expectations revolve around personal attributes. As mentioned, a positive attitude is extremely important. However, employer expectations also include attributes such as dependability, honesty, loyalty, cooperation, enthusiasm, motivated, an ability to follow directions, initiative, and acceptance of criticism.

Dependability

High on the list of traits an employer appreciates is dependability. ***Dependability*** means worthy of trust. Dependability also means the employee can be counted on to do a job as instructed. Consistent performance is a positive personal attribute.

Employers expect workers to be on the job every day and to report to work on time. Many people have lost their jobs because of failure to report to work or chronic tardiness. Don't miss work if at all possible. Call well in advance if sickness prevents you from going to work. An employer cannot keep someone on staff that is not dependable.

Honesty

Employers expect honesty from their employees. ***Honesty*** reflects truthfulness, integrity and fairness. It is broad ranging in the workplace. Lying to others at work is dishonest. Stealing materials from the company is dishonest. One of the major issues in terms of honesty is stealing time. Taking long breaks or not being on task is essentially stealing from the employer. Another dishonest practice would be calling in sick when illness is not a problem.

2–3. Cooperation means working with others for the common purpose. (Courtesy, Agricultural Research Service, USDA, Scott Bauer)

Loyalty

Loyalty or faithfulness to the company is expected of employees. Loyalty can be expressed when speaking to customers, co-workers, and friends. Employees should refrain from being critical of the company. After all, the company is providing an employment opportunity and a paycheck.

Cooperation

Cooperation means working with everyone for the common purpose. Acts of cooperation are many. Employees might be asked to do unpopular tasks. They should be done without complaint. Employees might be assigned to work with people they do not like. An effort should be made to get along in order to accomplish the task. Everyone should be willing to pitch in and do a fair share of the work. Willingness to be cooperative is often recognized and rewarded by employers.

Enthusiasm

Employees should go about their work with **enthusiasm**. Enthusiastic employees can be described as being lively, excited, involved, and interested in their work. Their positive attitude shows while they work.

2–4. Employers often support employees motivated to take classes.

Motivated

Employers look for employees that are **motivated**. Motivated employees take action. They have a desire to do a good job and to improve themselves. They are willing to learn. A willingness to learn company policies and company procedures is expected by employers. Motivated employees show an interest in learning jobs beyond their own. Employers often support employees in taking classes that will improve their skills and help them with their job. Those employees that expand their knowledge, position themselves for advancement in the profession.

Ability to Follow Directions

The ability to follow instructions is basically doing the work that is assigned the way it was told to be done. Employers expect employees to follow directions given to them. If there are questions regarding the instructions, the employee should request clarification.

The ability to follow directions can be applied to company policies. New employees receive instruction on company policies and are expected to adhere to them. For instance, company policies often mention the employee's appearance. Employees are typically asked to dress appropriately for the job. The reason for this is well dressed and groomed employees are a reflection of the company. It is important that employees follow company policies.

Initiative

Employees are sometimes confronted with situations for which they have no instructions. They can choose to sit around doing nothing or they can be self-starters. An employee who is honest, has learned about the job, and is enthusiastic will take it upon himself or herself to move on to another task without being told to do so. Moving on to a task without being instructed to do so shows ***initiative***. As an employee becomes more experienced, the employer will expect more in the way of initiative.

Acceptance of Criticism

There are times when employers will find fault with the work of the employee. Hopefully, the criticism is delivered in a positive way with the intent of improving the performance of the employee. Regardless, the employee should receive the criticism politely and make an effort to learn from the situation.

EMPLOYEE EXPECTATIONS

In return for quality work performance, employees can expect certain things from an employer. Expectations include wages, courteous treatment, training, evaluation of performance, honesty, and safe working conditions.

Wages

Compensation for work performed comes in form of wages. The employer should make the payments to employees in the form of a paycheck with deductions for income tax and social security. The employer must make separate payments to the government for social security. The employer may also provide other benefits to the employee such as medical insurance.

2–5. Employees might require training in certain procedures such as the safe application of pesticides. (Courtesy, Agricultural Research Service, USDA, Peggy Greb)

Courteous Treatment

Part of being a good leader is treating the people who work for you with respect. The employer should introduce new employees to the other workers. Company policies should be explained, as should any change in policies with time.

Training

An employee can expect the employer to provide the necessary training in order to accomplish the assigned jobs. Employees might need training in procedures. Most certainly, they will need training on the operation of equipment with which they are unfamiliar.

Evaluation of Performance

The employer should provide a periodic evaluation of the employee's work performance. Such an evaluation might include discussion on both an employee's strengths and weaknesses. The evaluation lets the employee know how their job performance can be improved. Evaluations often are used as the basis for promotions, raises, or dismissal from the job.

2–6. Employees can expect periodic performance evaluations.

Honesty

Employers should be honest with their employees. Compensation for work in terms of wages and benefits should be delivered as agreed upon.

Safe Working Conditions

Personal safety of the employee should not be jeopardized. Providing a safe working environment means training employees on safety procedures and removing hazards. Safety is a major issue in the greenhouse industry and more in depth discussion follows in this chapter.

WORKING WITH CO-WORKERS

Most jobs involve working with other people. Few people are successful in work if they cannot get along with others. One must develop skills to cooperate with people who have different personalities.

Why is cooperation so important? An employee is much happier if they have a good working relationship with their co-workers. The camaraderie enhances the working experience. In a cooperative working environment workers support one another. Greater productivity often results when employees get along.

It sometimes takes effort to get along with other employees. Not everyone is alike. People have different attitudes and beliefs. For two workers to have a good working relationship, they must accept their differences. Talking to each other is a good way to understand each other's beliefs and attitudes.

GREENHOUSE SAFETY

The nature of the work in a greenhouse can lead to personal injury. It is important for workers to be aware of common hazards they may encounter. Depending on the hazard, measures should be taken either to proceed with caution or to correct the problem. Some of the most common greenhouse hazards and safety procedures follow.

Wet areas—Wet and humid conditions in greenhouses create water puddles and algae buildup on walkways. The result is often slippery conditions that can lead to falls.

Recommended safety practices:

- Post warning signs indicating slick areas.
- Treat the slick surfaces to reduce algae growth.
- Periodically inspect the walkways.

2–7. Wet floors or aisles can promote algae growth and result in slippery conditions.

Cluttered aisles—Hoses, plants, carts and other items that clutter the aisles create tripping hazards.

Recommended safety practices:

- Check the aisles regularly to see that they are free of obstacles.
- Make sure hanging plants are secure and provide safe passage.

Faulty electrical wiring—Poor electrical wiring is a major cause of fires. Temporary wiring, improper splic-

2–8. Hoses lying in aisles can become tripping hazards.

2–9. Electrical outlets in the greenhouse should be well-grounded.

ing, and the excessive use of extension cords are the most common electrical problems in greenhouses.

Recommended safety practices:

- Have the fire department or a professional electrician inspect the greenhouse wiring.
- Use standard twist connectors to join wires.
- Only use extension cords with adequate capacity to carry the current.
- Never place extension cords under carpets where they can be damaged from traffic. Undetected damage can lead to smoldering.
- See that all joints are secured within a junction or receptacle box.
- Locate breakers and service boxes away from areas that will be wet.

Use of chemicals—Care must be taken when mixing and applying chemicals. Poisoning can occur through contact with skin and by inhaling fumes.

Recommended safety practices:

- Schedule the application of chemicals when temperatures are cool.
- Always read and follow instructions on the chemical label.
- Only allow employees trained in the use of chemicals to handle them.
- Wear safety gear when mixing and applying chemicals.

Power equipment—Both experienced workers and workers new to the industry are at risk of injury from pieces of power equipment. Inexperienced workers often are not aware of the dangers. They must be trained on the on the safety procedures. Experienced workers become familiar with procedures and can become careless. They often require re-training on safety practices.

Power equipment presents hazards that should be addressed. Some common danger zones and recommended safety practices include:

- Fans—Fan blades, flywheels, and pulley systems should be guarded. The guards should have small openings that prevent entry of fingers and hands.
- Potting machines—Drill heads on potting machines are dangerous and should be guarded.
- Soil mixers—Soil mixers should have safeguards that prevent workers from falling into the hopper.
- Conveyors—Automated conveyor systems have become increasingly popular. These systems offer a particular danger where the conveyor belt and end roller meet. These points need to be guarded so the machinery cannot grab clothing and fingers.
- Tractors—The power take off (PTO) must be guarded as it can grab limbs and clothing and cause serious injury. Also, tractors are to be ridden only by the operator of the vehicle. Additional riders should be prohibited.
- Forklifts—Forklift trucks are among the most dangerous pieces of power equipment used in the greenhouse industry. As a result, employers must provide forklift training to employees regarding the operation of the vehicle.

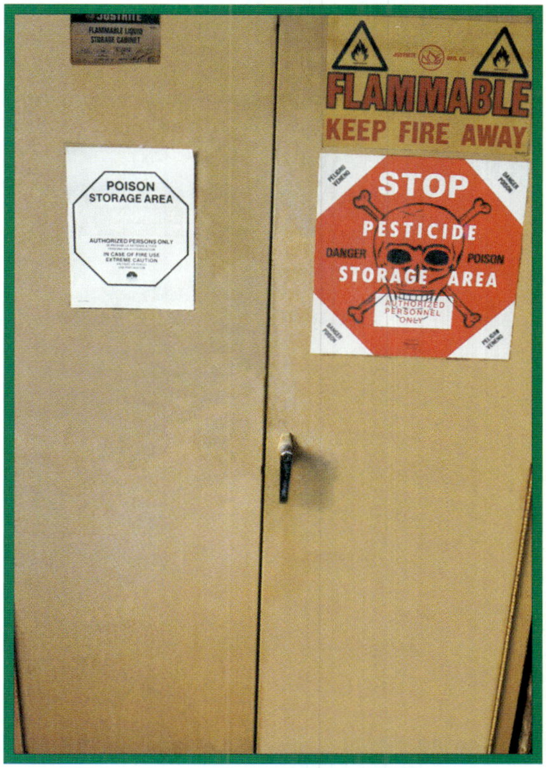

2–10. Chemicals must be stored safely.

2–11. Students gain real-world experiences through Supervised Agricultural Experience Programs. (Courtesy, The National FFA Organization)

SUPERVISED EXPERIENCE

Valuable work experiences can be obtained through ***supervised agricultural experience programs (SAEP)***. SAEs provide opportunities to apply classroom instruction through real-world experiences. Through SAEs students have an opportunity to explore the greenhouse industry.

Students practice specific job skills in one of four ways. They can have exploratory, placement, ownership, or research and experimentation projects. Common greenhouse oriented SAEs are placement or employment with a local business and ownership of a greenhouse operation.

The Agriculture/Horticulture instructor, the parents, and the employers provide supervision of the experience. These people are available to help plan a student's project and provide guidance as the experience progresses.

It is common for students to begin with a fairly simple SAE. As the student learns more the project may become more advanced. Student recognition for SAEs can be gained through the FFA Organization.

STUDENT YOUTH ORGANIZATIONS

THE NATIONAL FFA ORGANIZATION

The premier student organization for young people interested in the greenhouse industry is the ***National FFA Organization***. In fact, the FFA Organization is designed for students enrolled in Agriculture/Horticulture and is an integral part of the classroom instruction. The National FFA Organization is the largest student organization in the United States.

The focus of the FFA Organization is threefold. It is to promote premier leadership, personal development, and career success. ***Leadership*** is the ability to influence other people to meet individual or group goals. ***Personal growth*** deals with social skills, citizenship, and topics associated with self-esteem. ***Career success*** involves those things that will help a person succeed in the world of work. The employability skills mentioned in this chapter are addressed through the FFA Organization.

Many exciting experiences are possible through the FFA. Students can travel and meet new people. Students can hone their greenhouse skills by participating in floriculture and horticulture career development events. A student can also receive awards at chapter, regional, state, and national levels based on his/her proficiency in greenhouse management.

THE NATIONAL JUNIOR HORTICULTURAL ASSOCIATION

Another organization for young people interested in the greenhouse industry is the **National Junior Horticultural Association (NJHA)**. NJHA projects and activities are for youth 22 years of age and under. Membership in the NJHA is achieved by enrollment or participation in any NJHA project or activity.

The NJHA program focus is represented by the acronym HYCEL, which stands for Horticulture, Youth, Careers, Education, and Leadership. NJHA projects and activities are designed to work with existing programs, such as, but not limited to 4-H, FFA, scouts, horticulture/agriculture education, and youth gardening groups. Projects and activities are also designed for independent youth who participate directly in NJHA.

2–12. Students can test their greenhouse skills through FFA career development events. (Courtesy, The National FFA Organization)

REVIEWING

MAIN IDEAS

Employers have expectations of a worker. A positive attitude desired. Attitude can be the greatest factor in job success. A person is generally considered to have a positive attitude or a negative one. Some other expectations include dependability, honesty, loyalty, cooperation, enthusiasm, a desire to learn, an ability to follow directions, initiative, and acceptance of criticism.

Employees can expect certain things from an employer. Employee expectations include wages, courteous treatment, training, evaluation of performance, honesty, and safe working conditions.

Employees must cooperate with co-workers. An employee is much happier if they have a good working relationship with their co-workers. Greater productivity often results when employees get along.

It is important for workers to be aware of common hazards they may encounter in a greenhouse. Some of the most common greenhouse hazards include wet areas, cluttered aisles, faulty electrical wiring, chemical use, and power equipment. Both experienced workers and workers new to the industry must be trained on and follow safety procedures.

Valuable work experiences can be obtained through supervised agricultural experience programs (SAEP). SAEs give students an opportunity to explore the greenhouse industry. The opportunities are usually those that cannot be provided within the classroom setting. Students practice specific job skills through employment or by overseeing their own greenhouse operation.

The premier student organization for young people interested in the greenhouse industry is the National FFA Organization. The FFA Organization is for students enrolled in Agriculture/Horticulture and is an integral part of the classroom instruction. It promotes premier leadership, personal development, and career success. Another organization for young people interested in the greenhouse industry is the National Junior Horticultural Association (NJHA).

QUESTIONS

Answer the following questions. Use correct spelling and complete sentences.

1. What are some adjectives to describe positive and negative attitudes?
2. What is dependability?
3. How can honesty be defined?
4. Why is cooperation important?
5. What does it mean to show initiative?
6. What can the employee expect from the employer?
7. What are some major safety hazards found in greenhouses?
8. What are six safety hazards associated with power equipment?
9. How does a SAEP help prepare someone for greenhouse careers?
10. What are some ways the National FFA Organization can help prepare someone for greenhouse careers?

EVALUATING

Match the term with the correct definition. Write a letter by the term in the blank provided.

a. attitude
b. cooperation
c. dependability
d. enthusiasm
e. honesty
f. initiative
g. leadership
h. loyalty
i. National FFA Organization
j. supervised agricultural experience (SAE)

______1. Moving on to a task without being instructed to do so.

______2. Provide opportunities to apply classroom instruction through real-world experiences.

______3. The premier student organization for young people interested in the greenhouse industry.

______4. Employees described as being lively, excited, involved, and interested in their work are said to have this.

______5. Faithfulness to the company.

______6. Working with everyone for the common purpose.

______7. One's outlook on life.

______8. Reflects truthfulness, integrity and fairness.

______9. The ability to influence other people to meet individual or group goals.

_____10. Worthy of trust.

EXPLORING

1. Examine your own attitude. List the traits that would contribute to a positive attitude. Identify traits that are associated to a negative attitude and list those. Reflect upon your results. Make a conscious effort to behave in a positive way when tempted to act negatively. Practice those things that contribute to a positive attitude, such as smiling easily and being cheerful. Find life exciting and enjoyable. Avoid complaints and criticism. Respect the opinions of others, and accept responsibility for mistakes.

2. When a guest speaker appears in class or if a field trip is taken to a local business, ask the employer to discuss employment problems common to young employees. Follow up by asking what they expect in their employees.

CHAPTER 3

The Greenhouse Business

OBJECTIVES

This chapter has the following objectives:

1. Identify basic business principles for starting a greenhouse operation.
2. Describe considerations involved in locating a greenhouse.
3. Describe basic management of a greenhouse production enterprise.
4. Recognize the importance of marketing and promotion.
5. Discuss packaging and transporting greenhouse crops.

TERMS

acclimatization
advertising
business plan
corporation
fictitious name statement
image
inventory
marketing
organizational charts
partnership
sales
sole proprietorship

3–1. Having knowledge in producing greenhouse crops is seldom enough to succeed in establishing and managing a greenhouse business. (Courtesy, The National FFA Organization)

THE ABILITY to grow beautiful plants is truly a talent. For some it is a labor of love. Not surprisingly, many people with an interest in plant production see career opportunities in the greenhouse industry. They can earn a living doing something they enjoy. Fortunately, there are many opportunities in the industry. One option for them is to start a greenhouse enterprise.

Having the knowledge to produce outstanding greenhouse crops is great. Yet, horticulture knowledge alone is seldom enough to succeed in establishing and managing a greenhouse business. An understanding of business principles is necessary. Skills in financial management, marketing, and laws related to business operations are as important as production.

FUNDAMENTALS FOR OPENING A GREENHOUSE BUSINESS

Where does one begin in starting a greenhouse business? The basics would be a good start. Insight on six fundamentals can determine the level of success in starting a greenhouse business.

1. Before getting started talk to people with experience in starting a new business. Most people in the industry would readily provide advice. Learn from them things that should be done and those things that work. At the same time learn about things to avoid.
2. Secondly, identify important contacts. These contacts might be called upon frequently. Keep their names, addresses and telephone numbers close at hand. Key contacts might include:
   - other greenhouse business people,
   - bankers to help with financial advice,
   - lawyers to interpret laws and assist with legal documents,
   - accounting professionals to help with a business plan, financial records and taxes,
   - public relations/advertising professionals to provide promotional direction,
   - professional greenhouse associations for networking purposes and publications,
   - computer technicians to help with computer hardware and software systems,

3–2. Computers are essential business tools today and contact with a computer technician to assist with computer hardware and software systems is considered important.

- state university extension specialists for technical advice,
- vendors of supplies and equipment
- government agencies in charge of licensing and permits

3. Third, be sure to have a competitive edge. One or more unique aspects of the business could provide such a competitive edge. For example, a business location convenient to customers could improve sales. A location that makes transporting greenhouse products easier could provide the competitive edge.

   Find a market niche that could provide an edge. Produce a crop no other greenhouse in the area is producing. Produce an especially high quality product. Provide a friendly, polite, customer-oriented service unique to the area. Consider selling goods at lower prices to attract business. However, prices should never be so low as to put the company out of business and or to harm the industry.

4. It costs money to start up a business. Because of start up costs, it is common to start with a small enterprise and expand as sales increase. Relatives or family friends quite often provide money or financial backing. Loans are the most common financial assistance.

   If a loan is sought from a lending institution be prepared to provide a business plan. A ***business plan*** is an organizational business tool to state the business' purpose and goals and estimate the financing needed to fund the enterprise. Included in the plan should be the business goals, the business description, and the products provided. Needs in terms of structures and equipment, land and utilities, and the amount of money to get started should be stated. Also include the anticipated

3–3. Polite customer service as given by this salesperson could provide a competitive edge.

expenses of the business, collateral to secure the loan, a plan for repayment of the loan, and the legal business structure to be used. A business plan might include the following sections:

A Business Plan

Title Page — Logo
Company name and address
Owners' names and telephone numbers
Table of Contents

Statement of Purpose — 1) Business goals
2) Purpose of this document
*Application for financing
*Give summary of specific capital needs

Business Data

Market and Competition Analysis

Description of the Business — Type of greenhouse business
Business niche
Products and services

Business Operation — A description of operation from product procurement to production, sales, and delivery

Location — Give a description of the site
Give building specifications
List traffic-flow count
For a new location, identify at least two location options

Marketing Plan — The marketing plan will describe the initial promotion and advertising strategies as well as the long range plans for advertising and promotion

Future Change Strategies — List the long-range plans for growth and expansion

Management Team

Operating Personnel

Venture Development Schedule — For a new business, show a 12-month time frame of the business from its conception to its opening

Proposed Use of Requested Capital

Critical Risks and Problems — List the difficulties and problems that may occur after the business is open and the plans to overcome them

Summary

Financial Data (The prospective owner will need to submit a personal financial statement.)

Capital Needs and Sources — List the source of all capital, including the owner's funding and other loans and investments.

Capital Equipment List — List all equipment needs, including computer systems, vehicles, office furniture, and non-inventory items.

Balance Sheet at the Start of the Business

Break-even Analysis

Income Forecast

Cash-Flow Analysis

Balance Sheet — Three year summary and three yearly forecasts

Loan Proposal

---

5. Licenses and permits are required to operate a greenhouse business. Five licenses might apply to a greenhouse business. They are new business, employer, resale, professional, and special use. The required licenses are regulated at the federal, state, county, and local levels. State, county, and local regulations vary widely. Thoroughly check with officials to determine requirements.

New Business Licenses

New business licenses include a fictitious name statement, any operating licenses required by the state, county, or local governments, and any local building permits. A ***fictitious name statement*** is a document that must be filed with the county to register the business name and prevent another business from using that name in the county. The greenhouse company is required to pay a filing fee. The county clerk's office and the local city hall provide information on zoning laws, occupancy permits, and fire code compliance.

Employer Licenses

Greenhouse companies with employees must have a Federal Employer Identification Number (FEIN). Some states also require a State Employer Identification Number. Both the Internal Revenue Service and the Social Security Administration use the FEIN to identify the greenhouse business. If the greenhouse is a sole proprietorship with no employees, the owner's social security number is used instead of a FEIN to identify the business. Contact the Internal Revenue Service to apply for a FEIN and to request federal tax forms required of a new business.

A greenhouse company with one or more employees must comply with Occupational Safety and Health administration (OSHA) requirements. The latest regulations are available from the Department of Labor in Washington, D.C.

Resale Licenses

A retail greenhouse business must have a resale license or permit (also called a tax number or seller's permit). This registers the business as a seller, enables it to collect sales tax, and allows it to buy goods for resale from a wholesaler without paying tax. This license does not apply to states that do not collect tax. A tax number or resale license is applied for with the state's Department of Revenue. Once the permit is received, the greenhouse must forward sales tax to the state on a monthly, quarterly, or yearly basis, depending on the state. Check with the local county clerk's office and city hall to determine if municipal or county resale licenses are required.

Professional Licenses

Professional licensing for greenhouse companies is not mandatory in most states. The regulations should however be researched with the state Department of Agriculture before proceeding with financing and a new business.

Special Use Licenses

A common special use license is a pesticide applicator's license. Check with local and state agencies for the details. The pesticide applicator's license is administered through the state Department of Agriculture.

6. Finally, it is very beneficial to secure some sale orders before starting up. This provides some financial security. Making advanced sales is easier accomplished with a wholesale business than retail. Whether the company is to be wholesale, retail, wholesale/retail, or mail order, it is important to inform potential customers of the product to be produced before opening for business.

BUSINESS STRUCTURES

As mentioned, a legal business structure must be determined. There are several options from which to choose for the legal structure of the business. They are sole proprietorship, partnership, and corporation.

SOLE PROPRIETORSHIP

The assets and liabilities within a ***sole proprietorship*** belong to the owner or owners. It is the easiest of the three business structures to set up. The owner makes decisions. In turn, the owner receives all profits. Taxes on the profits are paid on the owner's personal income tax. Expenses and losses may be deducted from the owner's income tax.

On the other hand, the owner is personally responsible in lawsuits and for debts. Also, the owner must manage all aspects of the business. Management is likely without significant input from other individuals.

PARTNERSHIP

A ***partnership*** consists of a number of individuals who have agreed to share the responsibilities of the business. A written agreement is drawn up between the individual partners. Agreements outline money and other contributions by the partners; salaries; division of profits; duties, responsibilities, restrictions, and the authority level of each partner. There should also be provisions for admitting new partners; distribution of assets upon ter-

3–4. The president of this greenhouse corporation oversees the day-to-day business operations.

mination of the partnership; death or disability of a partner; and in the event partners disagree.

Partnerships are fairly easy to establish. They have the advantage of a greater pool of talent. Plus, additional people involved can provide greater resources of capital. Tax advantages are the same as those for a proprietorship.

Partnerships do have some disadvantages. Shared ownership can create problems, and legal disputes can arise with the disagreement of partners. Personal liability for each partner is unlimited. Also, should a partner die or leave the partnership, the partnership is consequentially dissolved.

CORPORATION

A **corporation** is a legal association of individuals. A corporation exists with powers and liabilities (i.e. money owed, debts) independent of its individual members. A board of directors is selected to guide the company. Officers such as the chairman, president, chief financial officer oversee the day to day operations of the company. Responsibilities of the corporation's board of directors and officers must be defined in a statement.

Capital can be raised through the sale of stock. Those who buy stock are stockholders. They are investing in the company. Ownership is transferred with the sale of stock from one individual to another. Stockholders are not personally liable for the business. Plus, the corporation can continue with the death of a stockholder.

There are some disadvantages to a corporation. It is a complex business structure, and as such, requires more effort in maintaining records, filing reports, and holding formal meetings. Both corporate profits and dividends to individuals are taxed.

LOCATING A GREENHOUSE

Location, Location, Location, a phrase common in real estate cannot be underestimated. Start off right by locating the greenhouse business where there is a good chance for success. Carefully consider a number of factors before deciding on a location.

1. Grow something people in the service area will buy. Most cities in the United States receive cut flowers from throughout the world. So potted plants and bedding plants would be the logical choice of most greenhouse operations.
2. The decision as to whether the business will be retail or wholesale will influence what crops will be grown. If retail, the greenhouses are usually in conjunction

with a florist shop or garden center. If wholesale, determine the market within a 20–30 mile range, less within a metropolitan area.

3. Wholesale operations should be a minimum of 25,000 square feet. Purchase 4 to 5 times as much land to allow room for expansion. Wholesale operations need not be on a major road. Returns on three major crops (for example; poinsettias, Easter lilies and bedding plants) a year need to cover business expenses and provide a profit.

3–5. This company boasts a greenhouse range with a garden center.

It is very helpful to calculate the possible returns per square foot. Deduct salaries and expenses from the total estimate to arrive at the profit.

1 pinched poinsettia, wholesale price	$5.00
2 Easter lilies (8 × 8 spacing)	$10.00
8 Geraniums	$10.00
Total return per square foot	$25.00

4. Check the zoning ordinances to determine tax rates. In addition, get an idea how the surrounding area is to be developed. Residential areas carry higher taxes. Industrial areas can produce harmful pollution.

5. Study the environmental features of the land. Investigate prevailing winds that can influence both the costs of heating and the movement of air pollution. The number of clear days verses cloudy days during the year can have a significant impact on production.

6. One of the most important considerations is the water supply. The water source should be low in salts, pollutants, dissolved minerals, and have a neutral pH. It should also be readily available and inexpensive.

7. An affordable labor supply is needed for both skilled and unskilled workers.

8. Transportation of goods is critical. Retail operations must be easily accessible from a major thoroughfare.

9. Consider the availability and cost of energy to heat and cool the greenhouse.
10. Consider the orientation of the greenhouse on the property. In general, the ridge of the greenhouse should run in an east-west direction to reduce shadows caused by the structure. This has greater importance if the greenhouse location is above the 40th latitude. Heating costs can also be reduced if the end of the greenhouse, which can be insulated, faces the direction of prevailing winter winds.

3–6. Many considerations go into locating a greenhouse. (Courtesy, Bell Nursery, Maryland)

MANAGING THE GREENHOUSE BUSINESS

Once up and running, the greenhouse business needs to be managed. Managing a greenhouse business requires skill in organizing and controlling the company. This can be accomplished by carefully coordinating the production, marketing, and financial segments of the business. The different segments must work together.

How does one go about coordinating the three major segments of a company? Smooth coordination hinges on three business functions. They are planning, delegation, and control.

PLANNING

Planning is essential. Achievable business goals must be set. Employees should understand the goals and be motivated to work towards achieving them. Progress is then measured to determine the degree of success.

It is advisable to develop short term and long-term plans. Both plans would include goals for production, marketing strategies, and financial plans. Short-term plans tend to be detailed as they typically involve one year of operation. Since it is more difficult to see into the future, five-year plans are less detailed. As the time period for short term plans closes, the accomplishment are evaluated and the results are used to develop the next year's plan.

Organizational charts are commonly used as diagrams that clarify responsibilities and lines of command. Organizational charts often are simple sketches. An organizational chart typically begins with the board of directors, president, or owner. The various departments branch out from there. Lines connecting the different workers indicate who reports to who.

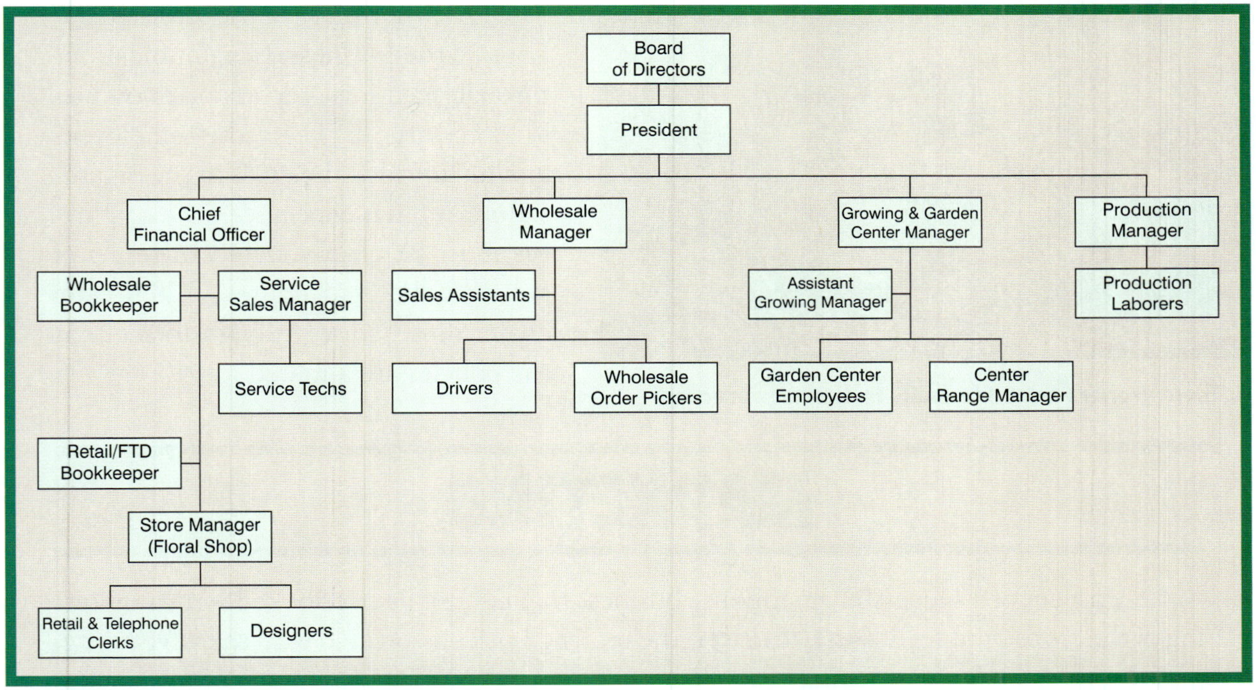

3–7. Shown is an organizational chart for a company that operates a wholesale greenhouse, retail florist, and garden center.

DELEGATION

Initially, the owner of the business might oversee all phases of the greenhouse operation. As the business grows, the owner might find that he or she simply cannot tend to all phases of the business. The owner must delegate some the work. Employees and family members must be given responsibility for different aspects of the operation. The details are left to them. However, they report to the owner.

Additional growth might warrant the formation of departments. Growers may be given charge of the production and taking inventory. ***Inventory*** is a detailed list of the number, quality, and value of plants and materials. Sales personnel oversee marketing. The business manager would handle the financial books. With an increase in size of the business and the number of departments, management of the business can become more challenging. Ironically, the owner might find himself/herself far from growing plants, which is what attracted him/her to the industry in the first place.

3–8. These growers meet informally to discuss production.

CONTROL

The owner must control the operations of the business. Communications through meetings, reports, memos, etc. is necessary to keep the employees moving towards the goals of the organization. What is expected of the employees must be clearly stated in terms of policies and procedures. Then, the owner must follow through to see that the policies are adhered to.

MARKETING

One aspect of a successful greenhouse business that cannot be overlooked is the marketing of the crops produced. ***Marketing*** is the total activities associated with the advertising, promoting, selling, and shipping of products and services. Marketing programs focus on the consumer.

IMAGE

The marketing approach begins with image. Although hard to define, ***image*** can be thought of as the impression people have of a business. Customers are influenced, positively of negatively, by the practices of a business. Companies that adopt the philosophy "the customer is always right" tend to be successful in building a positive image. That is because emphasis is placed on the customer.

It takes time and continual effort to develop a positive image. A good analogy is the saying that a person can spend a lifetime building a good reputation, but it only takes one lapse in judgment to destroy a good reputation. What are some things a business can do to build a positive image?

1. The appearance of the place of business leaves a lasting impression. Therefore, potential customers should find the appearance of the facility inviting. The interior and exterior should be well lit and clean. Products should be well organized. The workers should also be well groomed and dressed appropriately.
2. The quality of materials and services provided by the business greatly impacts its image. A business should strive for a yes answer to questions such as the following. Are the products of consistent high quality? Are the products delivered on time? Are the prices fair, while providing a profit for the business? Does a guarantee accompany the product? Are the hours of operation convenient for the customers?

3–9. Employees who treat customers with courtesy and respect contribute to a positive business image.

3. Business employees at all levels must treat customers with courtesy and respect. Employees with positive attitudes and effective communication skills contribute to building the image of the business. Studies have shown that the majority of customers are lost from a business because of discourteous treatment or because their complaints were not handled to their satisfaction. Further damage to the business image is done as these people tell their friends and family of the poor service.

4. ***Advertising***, efforts to encourage sales, creates an image. Advertising involves many different techniques. Some methods include newspaper ads, yellow pages, signs, direct mail, radio and television spots, and displays. The styles

3–10. This poinsettia was forced to flower in September as a marketing technique to show potential customers a sample.

used in the ads and frequency of use tells the customer about the business. Thought must be given as to the type of image the business wants portray with advertising.

A set of measurable goals should be established with an advertising plan. This will help in preventing an investment in advertising that may fail to produce the desired customer response.

5. ***Sales*** or the exchange of a product for money is what keeps a greenhouse business in business. The people that handle sales of the products carry the responsibility of maintaining the company's image. They can project a positive image by being courteous, polite, and professional. Handling customer orders efficiently, promptly returning calls, and assisting the customer with complaints can earn trust and customer loyalty.

3–11. This Maryland based company promotes its image on the side of its delivery truck. (Courtesy, Bell Nursery)

HANDLING AND SHIPPING CROPS

Responsibility for the greenhouse crop remains with the greenhouse company until the product is delivered to the customer. Greenhouse crops are perishable. That is they will be damaged or die if not handled carefully. Therefore, great care should be taken to see that crops reach the market in as good of condition as possible. Delivery should also be done quickly to reduce stress on the crop.

Good growers acclimatize the plants prior to shipping. ***Acclimatization*** is a process in which a plant becomes accustomed to a new environment. It involves adjusting temperatures, watering, and fertilizing practices prior to shipping. Details for specific crops can be found in production chapters of this book.

Before shipping a grower should do several things. The plants should be watered. The plants should be protected with plant sleeves and/or cardboard boxes. Removal of plants from the greenhouse, packaging them and loading the plants onto trucks should take place in less than a day.

3–12. This wholesale order picker sleeves a plant in preparation for delivery.

Cold weather presents additional concerns. Crops should be loaded onto trucks inside the headhouse. The trucks should maintain a minimum temperature of 40°F for most plants and 50°F degrees for cold sensitive crops.

REVIEWING

MAIN IDEAS

Six fundamentals can determine the level of success in starting a greenhouse business. Talk to people with experience in starting a new business. Identify important contacts. Be sure to have a unique feature that provides a competitive edge. Obtain money to start the business from relatives, family friends, or financial institutions. Check with officials for licenses and permits required to operate a greenhouse business. Secure some sale orders before starting up.

Determine the legal business structure to be used. Three options include sole proprietorship, partnership, and corporation. The assets and liabilities within a sole proprietorship belong to the owner or owners. It is the easiest of the three business structures to set up. Partnerships consist of a number of individuals who have agreed to share the responsibilities of the business. A corporation is an association of individuals and is created by law. Capital can be raised through the sale of stock.

Consider a number of factors before deciding on a location for a greenhouse. Some considerations include the market for crops, the region covered, the size of the operation, zoning ordinances, environmental features of the land, the water supply, labor supply, accessibility by road, energy costs, and orientation of the greenhouse.

Managing a greenhouse business requires skill in organizing and controlling the company. This can be accomplished by carefully coordinating the production, marketing, and financial segments of the business. Smooth coordination hinges on three business functions, planning, delegation, and control.

Marketing is the total activities associated with the advertising, promoting, selling, and shipping of products and services. Marketing programs focus on the consumer. The marketing approach begins with image. Image can be thought of as the impression people have of a business. The appearance of the place of business, the quality of materials and services provided, employee attitudes and communication skills, advertising, and the people that handle sales create the company's image.

Greenhouse crops remain the responsibility of the greenhouse company until the product is delivered to the customer. Since crops are perishable, great care should be taken during handling, storage, and shipping to see that crops reach the market in as good of condition as possible.

QUESTIONS

Answer the following questions. Use correct spelling and complete sentences.

1. What are six fundamentals that can be applied to starting a greenhouse business?
2. Who should be identified as contacts to possibly assist with the start-up process?
3. What is included in a business plan?
4. What are the three legal business structures?
5. What are ten factors to consider in locating a greenhouse?
6. What are three segments of a greenhouse business?
7. How is marketing defined?
8. What techniques might be involved in an advertising program?
9. How can the handling of sales affect the company's image?
10. Why is important that crops be handled carefully and shipped quickly?

EVALUATING

Match the term with the correct definition. Write a letter by the term in the blank provided.

a. acclimatization
b. advertising
c. business plan
d. corporation
e. image
f. marketing
g. partnership
h. sales
i. sole proprietorship
j. fictitious name statement

_____1. Business structure with a legal association of individuals.

_____2. The total activities associated with the advertising, promoting, selling, and shipping of products and services.

_____3. Efforts to encourage sales.

_____4. Business structure with assets and liabilities belonging to the owner.

_____5. An organizational business tool to state the business' purpose and goals and estimate the financing needed to fund the enterprise.

_____6. The exchange of a product for money.

_____7. Business structure with a number of individuals who have agreed to share the responsibilities of the business.

_____8. The impression people have of a business.

_____9. A process in which a plant becomes accustomed to a new environment.

_____10. A document that must be filed with the county to register the business name and prevent another business from using that name in the county.

EXPLORING

1. Set up a greenhouse business in the classroom. Use some of the business practices described in this chapter. Establish a business structure, produce crops, advertise, take orders, and deliver the products.

2. Maintain a greenhouse SAEP either at home or as an employee in a local greenhouse business.

3. Visit with a greenhouse owner. Ask questions that focus on what should be done to start a greenhouse business and what should be avoided. Record the answers received.

CHAPTER 4

Greenhouse Structures, Climate Control, and Automation

OBJECTIVES

This chapter is about greenhouse structures, climate control, and automation. It has the following objectives:

1. Identify greenhouse designs.
2. Identify and describe greenhouse glazing materials.
3. Recognize the advantages of retractable-roof greenhouses.
4. Explain environmental controls.
5. Describe methods of heating and cooling greenhouses.
6. Describe plant bench materials and spacing.
7. Describe automated systems.

TERMS

acclimatization
acrylic structured sheets
analog controls
automated seeders
automated transplanters
bench
capillary mats
computer controls
computerized environmental management
conservatory
ebb and flood
energy curtain
environmental controls
even-span greenhouse
fan and pad cooling system
fiberglass
fog system
glass
glazing
headhouse
hoop house
hot water system
infrared heat system
irrigation booms
lean-to greenhouse
polycarbonate structured sheets
polyethylene film
pot fillers
retractable-roof greenhouse
ridge-and-furrow greenhouse
rolling bench
short-day curtains
spaghetti tubes
steam heat
structured sheets
thermostats
tray mechanization
uneven-span greenhouse
unit heaters
vent

4–1. There are nearly 10,000 commercial greenhouse operations in the United States.

HAVE you ever had the opportunity to visit a greenhouse on a cold January day? It could be below freezing outside with a foot of snow on the ground, but the interior is full of life. Green growing plants, some in flower, are a welcome sight. The air feels warm and humid, as it might feel on a spring day. The atmosphere lifts the spirit, and thoughts of winter are left behind. It is no wonder people love visiting or working in greenhouses.

While greenhouse structures and the ability to control the climate within them create a pleasant atmosphere, they also contribute to a large industry. The facilities make the growing of plants possible throughout the year. Across the United States in 2001, greenhouse space for the production of floriculture crops total 541 million square feet. One usually does not have to travel far to see a greenhouse operation. The total of 8,579 greenhouse operations in 2001 also provided many employment opportunities.

GREENHOUSE STRUCTURES

The production of most floriculture crops greatly differs from other agricultural horticultural production. One factor that makes it so is the greenhouse. A greenhouse is a structure enclosed by a translucent material in which plants are grown. Translucent materials allow light, necessary for photosynthesis, to reach the plants inside. The material also serves as a barrier between the outside and inside environments. Greenhouse structures give the grower control of environmental factors within the structure. Growers can regulate the temperature, water, humidity and, to some extent, light.

4–2. Greenhouses give growers control of the growing environment.

TYPES OF GREENHOUSE STRUCTURES

Styles of greenhouses vary greatly. The material used, the shape of the greenhouse, and the load capacity contribute to the differences. However, there are four basic styles of greenhouse structures. They are lean-to, even span, uneven span, and ridge-and-furrow.

The ***lean-to greenhouse*** structure is attached to a building. The building provides support for the roof. If a lean-to structure is used for growing crops, it should be placed on the south side of the building.

Even-span greenhouse structures have roofs with an even pitch and an even width. These are single houses. ***Hoop houses***, so called because of the arching pipes used in the framework, are a type of low-cost, even-span greenhouse.

Uneven-span greenhouses have roofs with an unequal pitch and width. Use of the uneven span structure is limited to hillsides. Because the trend toward automated operations favors flat ground, few of these greenhouse structures are built today.

Ridge-and-furrow greenhouse structures consist of a number of greenhouses connected along the length of the houses. Shared side walls are eliminated, creating a large interior space. Fewer outside walls also result in lower heating costs. Gutters are installed where the individual houses connect to remove water. In cold climates, snow loads are a concern because the snow cannot slide off the roofs. One solution is to run heating pipes under the gutters to melt the snow.

Ridge-and-furrow greenhouse ranges are best oriented in a north-south fashion. This eliminates permanent shadows that would be cast by the gutters. Ridge-and-furrow houses are widely used in automated operations.

Modern greenhouse structures are being designed and built with higher gutters. That means the entire roof supported by the framework is higher. The higher structures provide better air circulation and more even temperatures. Higher structures also mean cooler summer temperatures at bench level.

4–3. A lean-to greenhouse.

4–4. An even-span greenhouse.

4–5. An example of an uneven-span greenhouse.

4–6. A ridge-and-furrow greenhouse.

Another advantage made possible in modern structures is the use of automated systems, such as energy curtains.

While crop production occurs in the greenhouse, many other tasks are performed in a headhouse. A ***headhouse*** is a structure attached to the greenhouse. It serves as a storage area for growing media, containers, fertilizers, and other materials. It is a work area. The potting and packaging of plants take place in the headhouse. In cold climates, headhouses often have overhead doors to allow trucks to enter. The trucks can then be loaded indoors without the risk of damaging the plants. Chemicals are stored in the headhouse in locked storage cabinets vented to the outside. Also, office space is often included in the headhouse so staff can conduct business.

4–7. This modern greenhouse has high gutters allowing the use of automated systems.

4–8. This headhouse includes material storage, restrooms, offices, a potting area, and plenty of room to package plants for shipping.

THE GREENHOUSE FRAMEWORK

The greenhouse structure consists of a framework that provides support for the covering material. Ideally, a framework should be strong while creating the least amount of shade possible. Narrow, light frameworks allow the maximum amount of light to reach the plants. The framework must be strong. It must be able to withstand snow loads in areas where it snows and hold the coverings under high wind conditions. Another important consideration is the

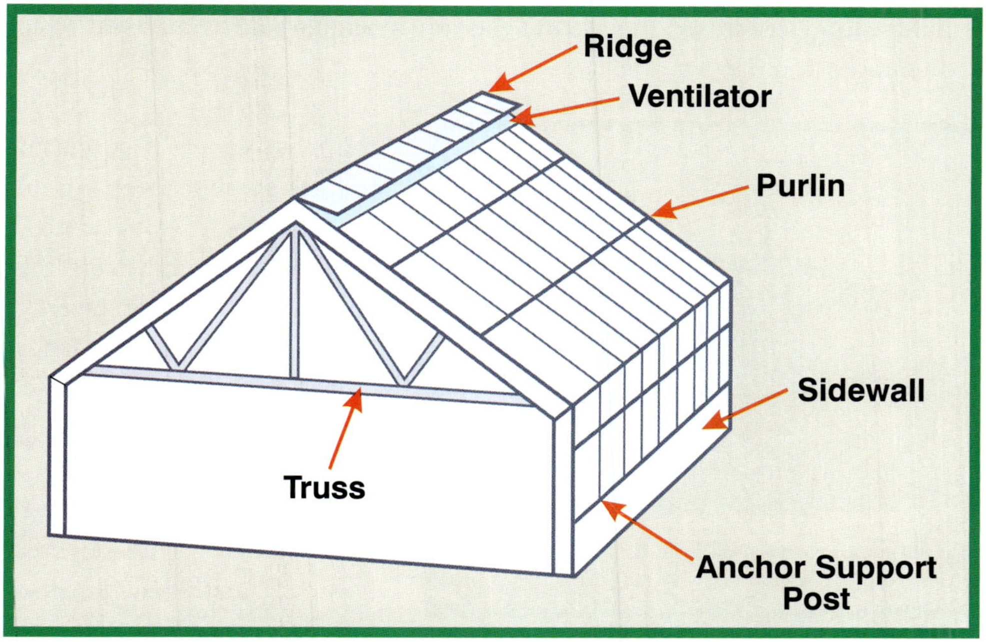

4–9. Structural components of the greenhouse framework.

4–10. The frame must be strong enough to withstand snow loads.

amount of maintenance the framing material requires. There is a great advantage to using framework materials that are maintenance free.

Aluminum and aluminum/steel combination frameworks are commonly used in modern greenhouses. Though somewhat costly, they are long lasting and maintenance free. Wood, angle iron, and galvanized steel are used as well. The disadvantages of these materials are wood and iron need to be painted, while galvanized treatments wear off.

COVERING (GLAZING) MATERIALS

The covering of a greenhouse is often referred to as the ***glazing***. There are several considerations in choosing a glazing material. Durability or the life of the material before it needs to be replaced is an important factor. Light transmission is important. The cost of the material needs to be considered. A fourth consideration is that of heat loss, which affects heating bills.

Polyethylene film, structured sheets, and glass are available glazing materials. Each material has advantages and disadvantages.

4–11. Polyethylene is used to cover these greenhouses.

Polyethylene (poly) film is the most widely used glazing material. Its flexible sheets, usually 6-mil in thickness, are stretched over the greenhouse framework. It is generally used in double layers with air blown between the layers to produce a 3- to 9-inch bubble-like covering. The air bubble effectively insulates the house. Heating costs of a double poly house are roughly one-third less than that of a glass-covered greenhouse. Polyethylene sheets are also inexpensive.

Poly is not very durable, however. That is because ultraviolet light breaks down the plastic. Improved poly sheets with ultraviolet light inhibitors last three to four years before they need to be replaced. One layer of poly cuts light transmission to 85 to 90 percent, and two layers of poly reduce light transmission to 70 to 75 percent. The lower light transmission limits the type of crop that can be produced, particularly in the North. Poly houses have a tight growing environment that doesn't "breathe." As a result, humidity levels can become quite high. Under humid conditions, a fungus disease of floriculture crops, *Botrytis*, can become a problem.

4–12. Condensation can be a problem with poly houses.

4–13. Structured sheets have gained in popularity. (Courtesy, CO-EX Corporation)

Structured sheets have grown quickly in popularity. Structured sheets are made from three materials. They are polycarbonates, acrylic, and fiberglass.

Polycarbonate structured sheets are the most widely used structured sheets. They are commonly manufactured with a "twin wall" held together by ribs. The resulting appearance is hollow tubes, side by side, running the length of the sheet. The tubing effect provides insulation and cuts heating costs. Polycarbonate sheets are treated with ultraviolet light inhibitors and are typically guaranteed for 10 years.

Polycarbonates have a good light transmission—80 percent. They resist hail damage. For that reason, polycarbonates are the product of choice in the hail belt stretching from Colorado to Texas. Polycarbonate sheets are also easy to work with and flexible. They can be bent around hoop greenhouses. Polycarbonates are fire resistant. A drawback is polycarbonate sheets are expensive.

Acrylic structured sheets are manufactured with twin walls in the same fashion as polycarbonate sheets. Therefore, they provide good insulation. They have a high light transmission, 86 percent, second only to glass. Acrylics last 8 to 10 years. Acrylics cost a bit more than polycarbonates. They are less flexible than polycarbonates, and they are more prone to hail damage.

Fiberglass was widely used in the 1960s and 1970s. Its use has dropped significantly since the introduction of polycarbonates and acrylics. Light transmission on new sheets is good, but it discolors after 7 to 10 years. Hail can cause some damage. Fiberglass is highly flammable, and it has poor insulating ability.

Glass is considered the best material for plant production. It provides high light transmission at up to 90 to 92 percent. The panes of glass overlap producing openings that allow air exchange. The initial cost of glass is expensive. It is very long lasting, provided it doesn't break. In fact, over the long term, it can be less costly than other coverings.

4–14. Glass is the best material for plant growth.

Unfortunately, glass is breakable. Greenhouse owners find the purchase of hail insurance and crop insurance necessary. Tempered glass, which is about double the strength of regular glass, is available. It withstands hail better, but it is more expensive. The increased strength allows for the use of larger panes. Supports for the glass are placed farther apart increasing light transmission. Also, heat loss is reduced with the larger panes. Still, glass has very poor insulation qualities.

Table 4–1. A Comparison of Glazing Materials

Glazing Material	Advantages/Disadvantages	Light Transmission	Lifespan (years)
Polyethylene (poly) film	Advantages: Inexpensive Easy to install Disadvantages: Not very durable Humidity levels can become quite high.	One layer 85–90 Two layers 70–75	1–3
Polycarbonate structured sheets	Advantages: Good light transmission Good insulation Resists hail damage Easy to work with Fire resistant Disadvantages: Expensive	80	10
Acrylic structured sheets	Advantages: Good insulation High light transmission Disadvantages: Less flexible than polycarbonates Expensive Prone to hail damage.	86	10–20
Glass	Advantages: High light transmission Glass overlap allows air exchange. Disadvantages: Breakable Initially expensive Heavy Very poor insulation qualities	Up to 90–92	25+

RETRACTABLE-ROOF GREENHOUSES

Technological advances provide another option for growers in terms of greenhouse structures. ***Retractable-roof greenhouse*** structures have roofs that can be opened and closed. They allow the grower to open the house to the elements when weather is favorable for plant growth and close the roof when the crops need protection. When open, crops are exposed to 90 percent sunshine and fresh air. The plants grow in a more natural environment. Retractable-

4–15. Retractable-roof greenhouse structures have roofs that can be opened and closed.

roof greenhouses are becoming more popular because they are a low-cost way to provide extra growing space.

Three main styles of retractable-roof greenhouses are on the market today. Flat roof houses are used for the purposes of shade and frost protection. Peaked roofs are designed for snow loads and keeping out rain. Sawtooth roof designs allow venting when other parts of the roof are closed. Polyethylene is used to cover the retractable-roof structure. When desired, sections of the poly are withdrawn and folded up by electric motors. The entire roof can be opened or closed in minutes.

Retractable-roof greenhouses are excellent for crops that prefer the outdoors, such as bedding plants, perennials, and field-grown cut flowers. Retractable-roof greenhouses also help plants become acclimated to the outdoors through exposure to the natural elements. ***Acclimatization*** is the process in which plants adjust to a different growing environment. Other advantages are fewer disease problems due to better ventilation, reduced irrigation because of rainfall, and more effective temperature control for DIF.

Depending on the location of the greenhouse operation, a retractable-roof greenhouse may or may not be used year-round. Some growers use retractable-roof greenhouses for poinsettia production in the fall. Northern growers leave the greenhouse idle in the winter months. In warmer areas of the South, however, the structure may be in use throughout the year.

Table 4–2. Advantages to Retractable-Roof Structures

- Crops are exposed to 90 percent sunshine.
- Crops are exposed to fresh air.
- Can be closed to protect plants.
- Crops grow more rapidly.
- Crops develop with a naturally compact growth habit.
- More effective temperature control for DIF.
- Less use of growth retardants.
- Fewer insect infestations.
- Fewer disease problems due to better ventilation.
- Less use of pesticides.
- Help plants become acclimated to the outdoors through exposure to the natural elements.
- Reduced irrigation because of rainfall.
- More profitable.

Greenhouse Benches

The level at which crops are grown in a greenhouse depends on the crop itself. The intended use of the greenhouse is another factor that is considered. A **conservatory** is a greenhouse designed for the display of plants. Conservatory plants are grown in beds on the floor to give a natural appearance. Since conservatories house tropical plants, including trees, they are the largest greenhouses built.

4–16. Conservatory plants are often grown in ground beds for a natural appearance. (Courtesy, *Greenhouse Product News*)

In production operations, plants might be grown on benches, on the floor, or in beds. Bedding plants and potted plants in most production greenhouses are grown on benches two to three feet off the ground. However, some production does occur on the floor of the greenhouses. Cut flowers, such as roses and carnations, are usually grown in beds within the greenhouse. This is because cut flowers are not moved during production. Care and harvesting of the cut flowers is also easier when the plants are grown lower.

4–17. Benches hold plants above the ground.

A **bench** is a structure that holds crops off the ground. The advantages of benches are numerous. To begin with, the plants are at a level that is comfortable for greenhouse employees. The raised benches allow water to drain from the containers. Also, it is easier to provide heat from below the plants.

Benches are manufactured from materials that can withstand the wet conditions in a greenhouse. They must also be strong enough to carry a lot of weight. Modern greenhouses use benches made from 13-gauge expanded galvanized steel, aluminum, and plastic. An alternative material for benches is rot-resistant wood, such as redwood or cedar. The recommended widths of benches are 3 feet when working from one aisle and 6 feet when working from both sides.

Benches are arranged depending on the use of the greenhouse. In conservatories and educational greenhouses, aisles are wide to allow the movement of groups of people. The benches are usually stationary, too. In production greenhouses, the goal is to use as much space as possible for the growth of crops. Therefore, aisles are kept narrow, often the minimum of 18 inches. ***Rolling benches*** maximize growing space in the greenhouse. Rolling benches rest on pipes. This allows each bench to be rolled from side to side. With this system, only enough space needs to be kept for one aisle. One person can easily roll a large bench to create an aisle.

4–18. Rolling benches maximize growing space in this greenhouse.

CLIMATE CONTROL

When one thinks of a climate, the temperature, humidity, precipitation, and the amount of sunshine often come to mind. Florida is considered to have a warm climate, while Alaska is considered to have a cold climate. People have no control of these large climates. However, greenhouse structures and climate control systems provide growers with the power to control temperature, water, humidity, and light within the greenhouse.

The greenhouse structure itself creates an enclosed environment. Yet, without the ability to manage the greenhouse climate, production of quality plants would not be possible. For example, quality of a crop is often determined by the ability to maintain temperatures within a few degrees. ***Environmental controls*** are devices used to automatically turn

4–19. Growers monitor growing conditions.

4–20. Thermostats are low cost and easy to install.

greenhouse systems on or off. Four types of environmental controls are associated with greenhouse use.

Thermostats are low cost and easy to install. There are two types, on-off and proportioning thermostats. On-off thermostats control fans, heaters, and vents with a change of temperature. Proportioning thermostats provide continuous control of systems with changes in temperature.

Analog controls use proportioning thermostats or electronic sensors to run amplifiers and electronic circuitry. They integrate the operations of heating and cooling equipment. The result is better performance than thermostats alone.

Computer controls utilize microprocessors. The microprocessors make

4–21. Computer controls utilize microprocessors to make complex judgments based on information from a number of sensors. (Courtesy, Ed Pryor Photographers Ltd.)

complex judgments based on information from a number of sensors. Those sensors might gather information on temperature, relative humidity, and sunlight.

Computerized environmental management systems are expensive, but they are accurate and offer the greatest range of uses. All pieces of automated equipment can be controlled together. For instance, vents may be opened or closed at the same time curtains are opened or closed. Computerized environmental management systems also can be programmed to provide different zones in the greenhouse with different climate conditions.

In recent years, computerized environmental management systems have made controlling the climate easier. The computers are sophisticated and they have become very affordable. Most growers are using computerized systems or are looking into adding computers to their operations.

4–22. A plant damaged by exposure to cold temperatures.

HEATING AND COOLING

Advanced growing schedules for floriculture crops involve pinpoint temperature control. Temperatures must vary between night and day as well as during different stages of the crop. Full control of temperature involves efficient cooling and heating systems. This is particularly true in the North.

During the day, cooling is a main concern. Radiant energy from the sun can quickly raise greenhouse temperatures. The heat can throw plant production off schedule. For example, potted chrysanthemum crops can be lost from heat stall. It can cause undesirable growth, such as Easter lilies stretching. In severe heat, plant tissues can be damaged. Excessive heat inside a greenhouse can be a problem even when it is cold outside.

The ability to heat a greenhouse is equally important. Greenhouses need additional heat on cold, cloudy days and at night. It is at those times when levels of radiant energy are low. On cold, partly sunny days it is common for the heating and ventilation system to work alternately. When sunny, the house may need to be cooled. When a cloud blocks the sun's rays, temperatures drop, and heat is required.

Heating Systems

Greenhouses are heated by several different methods. The trend is to use ***hot water systems***. Modern boilers are small and efficient. The water is heated in the boilers and pumped through pipes in the greenhouse. The pipes are most often placed under the benches or in the floor. Hot water systems have relatively low maintenance. The heat is delivered evenly. The temperature of the water can be adjusted as needed. Hot water heat systems can also be designed to provide separate temperatures to different houses.

4–23. Heating pipes placed under benches deliver heat under the crop.

Steam heat is used in greenhouses. Steam has been the standard method for heating cut flower operations. Large boilers bring water to a boil to produce steam. The steam then flows through pipes in the greenhouse. Steam heat is not as uniform as hot water heat nor is it as easy to adjust the temperature. One advantage to steam heat is the steam can be used to sterilize growing medium.

4–24. Water is heated in boilers to produce steam heat.

Small greenhouses and double-poly hoop houses are generally heated with ***unit heaters*** or hot air heaters. Air is heated within the unit and then blown by fans throughout the greenhouse. Unit heaters are less expensive to install than hot water systems. However, the long-term fuel costs for unit heaters are higher. Some growers prefer the heat to be delivered under the bench. Unit heaters deliver the heat above the crops. A common practice is to install unit heaters in large ranges as backup heaters for hot water or steam systems.

4–25. These unit heaters provide the necessary heat in this retractable-roof greenhouse.

Many growers choose to use polyethylene tubes to evenly distribute warm air. They generally run the length of the greenhouse, and they have holes along the side of the tube. The tubes are attached to fans associated with unit heaters. When the fans are operating, the tubes inflate. Air is blown through the holes in the tube. Polyethylene tubes promote an even circulation of air in the greenhouse. The result is uniform temperatures and fewer problems with foliar diseases.

Infrared heat systems produce heat energy that is absorbed by the plants, media, and benches. The heat then transfers to the air space around them. Therefore, infrared heat does not heat the air directly. The whole system is very similar to the way the Sun heats

4–26. Polyethylene tubes help circulate air for uniform heat and to reduce foliar diseases.

4–27. This photo shows propane-fired infrared heaters at the peak of the greenhouse.

Earth. High fuel efficiency of infrared heat results in low operating costs. Also, foliar diseases, such as *Botrytis*, appear to be less of a problem. Infrared heat systems are placed at the peak of a greenhouse so the energy can radiate throughout the house. For that reason, they function best in higher structures. On the down side, the required placement of the system eliminates the use of energy curtains.

Energy Curtains

Modern greenhouses are equipped with energy curtains. ***Energy curtains*** are automated systems utilizing fabrics that insulate the greenhouse at night and shade the crops during the day. The curtains are installed from gutter to gutter. With today's higher greenhouse structures, the curtains are 8 to 12 feet above the ground. Computerized

4–28. Energy curtains can be opened to allow maximum light or closed to shade the crops during the day.

4–29. Short-day curtains can be closed to darken a greenhouse for short-day crops like these poinsettias.

environmental management systems open and close the curtains based on preset light levels or temperatures.

Research has shown that energy curtains have a significant effect on greenhouse temperatures. The shade produced by curtains results in temperatures 8 to 10 degrees cooler than in full sun. The cooler temperatures reduce stress on the plants. Also, less water and fertilizer are required in the production of the crop. The curtains have an opposite effect when closed at night. Heat energy is contained within the greenhouse. The energy saved in terms of fuel costs ranges between 25 and 35 percent. For this reason, the use of energy curtains in colder regions has increased.

The same technology applies to the use of ***short-day curtains***. These curtains block light from reaching the crops. When closed for 12 to 13 hours a night, they produce a short-day (actually a long-night) effect. This ability to control light duration is critical in the timing of poinsettia, chrysanthemum, and kalanchoe crops. Two materials used for the curtains are a black polypropylene or an aluminum and polyethylene combination. The environmental control system opens and closes the curtain at prescribed times each day.

4–30. The number one cooling system used in greenhouses today is the fan and pad system.

Cooling and Ventilation Systems

The number one cooling system used in greenhouses today is the fan and pad cooling system. ***Fan and pad cooling systems*** are based on the evaporation of water. These systems involve a wall of cellulose or aspen pads on one end of the green-

4–31. Fog systems involve an atomizer that produces water vapor that cools the greenhouse. (Courtesy, Atomizing Systems, Inc.)

house. The pads are kept wet by a system of pumps and gutters that recirculate the water. On the other end of the greenhouse, there is a series of fans. The fans pull air through the pads and across the greenhouse. As the air passes through the pads, the water evaporates. The air is cooled as the water evaporates. This process can be compared to the cooling effect perspiration has on the skin. Fan and pad systems provide lower temperatures in the areas closest to the pads. The air warms as it crosses the greenhouse.

Another method of cooling involves a **fog system**, sometimes referred to as a mist system. Fog systems involve an atomizer that produces water vapor. The water vapor cools the house by flash evaporation. Fog systems increase the humidity within the house while they cool. One advantage of fog systems is that they effectively lower the temperature throughout the entire house. Fog systems are most often found in the southern states.

4–32. The roof of this modern greenhouse opens to provide ventilation and to increase light levels.

Natural cooling through ventilation is used where possible. **Vents** consist of panels that open and allow air exchange with the outside. Roof vents are effective. Hot air rises. As it does so, it escapes through the vent located at the peak of the house. Operating roof vents is much less expensive than fan and pad systems. Some greenhouse designs use both systems of cooling. Initial cooling is achieved by roof vents. However, if temperatures rise, the roof vents shut and the fan and pad system turns on.

OTHER AUTOMATED SYSTEMS

Many automated systems being used today involve handling of plants. Handling of plants is labor intensive. Growers typically handle millions of plants in a year. The labor required to propagate, pot, move, care for, harvest, and package the plants is costly. Cut flower production is the most labor intensive. Labor alone can account for 35 percent of the production costs. Bedding plant production also requires a great deal of labor. Fortunately, technological developments have assisted growers in performing some tasks.

Some of the technological developments are discussed here briefly.

POT FILLERS

One of the first automated greenhouse systems helped workers fill pots or flats with growing medium. The **pot fillers** fill pots with the medium. Workers feed empty pots to be filled into the mechanism. Pot filler machines fill the pots and lightly pack the medium. Another worker receives the pots for stacking or the pots move down a conveyor for planting.

4–33. Even with technological advances, floriculture production can be labor intensive.

4–34. These workers use a pot filling machine to fill pots with growing medium.

AUTOMATED SEEDERS

Automated seeders permit the sowing of hundreds of flats of bedding plant seeds. The seeders require few workers for operation. Larger producers of bedding plants rely on automated seeders.

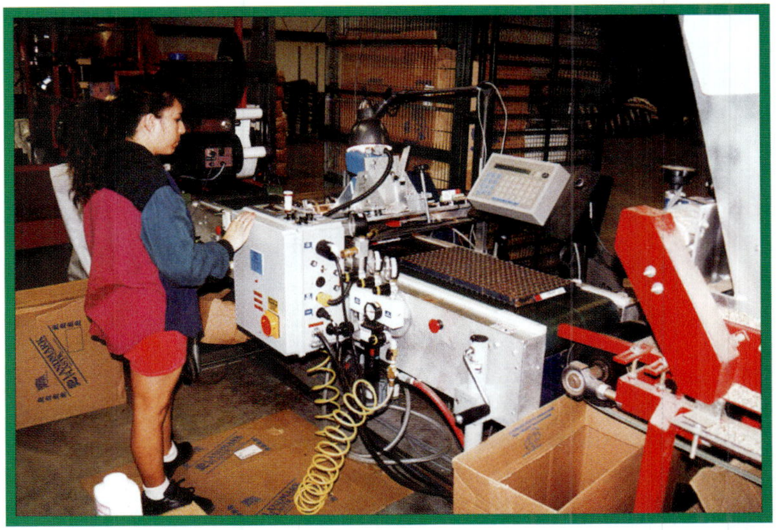

4–35. This person uses an automated seeder to sow thousands of flats of bedding plant seeds in a season.

AUTOMATED TRANSPLANTERS

Automated transplanters have become standard equipment in middle- and large-sized greenhouse operations. Small plants are removed from seed flats and planted into pots for finishing.

4–36. These people are receiving pots that an automated transplanter has planted with small plants from flats.

TRAY MECHANIZATION

One of the most labor-intensive activities in the greenhouse is the moving of plants to and from the growing area. Carts have been used for years. Conveyor belts are one of the most efficient ways to move plants in the greenhouse. Another transport method gaining popularity is ***tray mechanization***. Also known as Dutch trays or palletized benches, this system consists of benches that hold 100 or more pots. The trays are rolled on a series of rails from one area of the greenhouse operation to another. In more advanced operations, robots have replaced the rail system and move the trays to growing areas using bar codes.

4–37. Tray mechanization is a transport method gaining popularity.

AUTOMATED IRRIGATION SYSTEMS

Automated irrigation systems save labor. Some automated watering systems are ebb and flood, capillary mats, spaghetti tubes, and irrigation booms.

4–38. Bar codes were automatically placed on these pots at planting.

4–39. This ebb and flood system involves pumping nutrient solutions into the watertight benches at regular intervals.

Ebb and Flood—With an ebb and flood system, water is pumped into a watertight bench at regular intervals. The medium absorbs the water through the drain holes of the pots. Ebb and flood benches conserve water, provide accurate nutrient levels, and allow for optimum spacing of plants. Studies have shown ebb and flood systems to be 10 times as efficient as overhead watering.

Capillary mats—Capillary mats consist of porous mats laid on a bench and wetted. The medium in the pots absorbs the water by capillary action. This system is commonly used with plants, such as African violets, whose leaves can be damaged by overhead watering.

Spaghetti tubes—Spaghetti tubes involve tubing that provides water to individual pots. The end of each tube is fitted with an emitter that slowly discharges water.

Irrigation booms—Irrigation booms water large areas of bench space quickly making them useful for bedding plant production. Computers control the booms.

4–40. This grower programs a computer to prepare the desired concentration of nutrients in the tanks behind him.

4–41. This greenhouse has an ebb and flood system that delivers water to plants on a concrete floor.

4–42. These poinsettias are being watered with an irrigation boom that sweeps across the length of the greenhouse.

HAND AND POWER TOOLS

Greenhouse personnel rely on equipment and a variety of hand and power tools. Safe practices should always be followed. Tools should be kept clean so that they function properly and are less likely to transfer disease from one crop to another. Automated equipment needs to be regularly maintained and serviced. Tools should be stored in an organized manner after use.

4–43. Some hand tools used in greenhouse operations.

REVIEWING

MAIN IDEAS

Greenhouse structures are enclosed by translucent materials. They give growers of floriculture crops control of environmental conditions. Four basic greenhouse structures are lean-to, even span, uneven span, and ridge-and-furrow. A headhouse is a structure attached to the greenhouse. It serves as a work area and as a storage area.

The covering of a greenhouse is often referred to as the glazing. Polyethylene is the most widely used greenhouse covering. Structured sheets, now popular, are made of polycarbonates, acrylic, and fiberglass. They offer good light transmission, and some have excellent insulation properties. Glass is the best glazing material for the production of floriculture crops. However, it is expensive and breakable. Technological advances have provided retractable-roof greenhouse structures. Their roofs open and close allowing plants to be exposed to the elements when weather is favorable. Benches hold crops off the ground. Rolling benches maximize the growing space in a greenhouse.

Environmental controls are devices used to automatically turn greenhouse systems on or off. Four types available include thermostats, analog controls, computer controls, and computerized environmental management systems. Greenhouses are heated by the use of hot water, steam heat, unit heaters, and infrared heat systems. Energy curtains are automated systems used to insulate the greenhouse at night and shade crops during the day. Fan and pad systems, fogging systems, and vents cool greenhouses.

Automation has reduced the labor involved in the greenhouse industry. Major automated systems include pot fillers, automated seeders, automated transplanters, plant transport systems, and automated irrigation systems.

Equipment, hand tools, and power tools must be kept in good working order through proper maintenance practices, and they should be stored after use.

QUESTIONS

Answer the following questions using complete sentences and correct spelling.

1. Why are greenhouses so valuable to greenhouse growers?
2. How do the styles of greenhouses differ?
3. What are important considerations when selecting the greenhouse framing?
4. What are the advantages and disadvantages of the different glazing materials?
5. Why are retractable-roof greenhouses gaining in popularity?
6. What is used to control the greenhouse environment?
7. How are greenhouses heated?
8. What systems are used to cool greenhouses?
9. What are some technological advances in automation that have reduced labor?
10. How do the four major automated watering systems differ?

EVALUATING

Match the term with the correct definition. Write the letter of the term on the line provided.

a. automated seeders	e. acclimatization	i. unit heaters
b. fan and pad	f. ebb and flood	j. retractable-roof greenhouse
c. environmental controls	g. headhouse	k. rolling benches
d. polyethylene (poly)	h. ridge-and-furrow greenhouse	l. glass

_____1. The process in which plants adjust to a different growing environment.

_____2. Method of heating that involves the heating of air, which is blown by fans throughout the greenhouse.

_____3. Considered the best material for plant production as it provides the highest light transmission of any glazing material at 88 to 89 percent.

_____4. Structures have roofs that can be opened and closed.

_____5. Permit the sowing of hundreds of flats of bedding plant seeds.

_____6. Benches that rest on pipes for movement, thus maximizing the growing space in the greenhouse.

_____7. Structure attached to the greenhouse that serves as a storage area and a work area.

_____8. Number one cooling system used in greenhouses today. Cooling is based on the evaporation of water.

_____9. Automated system in which water is pumped into a watertight bench at regular intervals.

_____10. Structure consisting of a number of greenhouses connected together along the length of the houses.

_____11. Devices used to automatically turn greenhouse systems on or off.

_____12. The most widely used glazing material consisting of flexible sheets, usually 6-mil in thickness.

EXPLORING

1. Contact a greenhouse operation to arrange a visit. During the visit, note the construction materials used for the structure(s). Record the types of automated equipment used in production of the crops. Ask the grower to explain how the structure and equipment operate. Another objective might be to keep a list of conditions that could be hazards to the greenhouse workers.

2. Familiarize yourself with the high school greenhouse. Ask your instructor how it is designed to operate. What are some of the differences between the high school greenhouse and the larger commercial operations?

CHAPTER 5

Growing Media and Plant Nutrition

OBJECTIVES

This chapter is about growing media and plant nutrition. It has the following objectives:

1. Describe the qualities of a good growing medium.
2. Identify materials used as growing media.
3. Explain the differences between soil and a soilless medium.
4. Identify nutrients for plant growth and fertilizers that provide nutrients.
5. Describe recommended fertilization practices.
6. Describe the importance of pH on greenhouse crop production.

TERMS

aeration
available water
bark
bulk density
calcined clay
cation exchange capacity
coir
fertilizer
fertilizer injector system
growing medium
macronutrient
micronutrient
mineral soil
nutrient deficiency
organic matter
parts per million (ppm)
peat moss
perlite
pH
plant nutrition
plastic foam
porosity
primary macronutrients
rock wool
sand
secondary macronutrients
slow release fertilizers
soilless mix
soil pH
soluble salts
unavailable water
vermiculite
water-soluble fertilizers

5–1. The health of a plant is often determined by the growing medium.

HAVE you ever noticed a newborn baby? It is dependent on its parents for everything it needs. When the baby is hungry, it cries to be fed. When the baby is thirsty, it cries for something to drink. The baby lets it be known if it is experiencing other discomforts as well. Since the baby's sole means of communication is crying, parents often have to guess what the baby needs.

Plants are similar to babies in some ways. They have needs which, if not addressed, could result in death. Plants are dependent on the growing medium in which their roots grow. They draw moisture and nutrients from the growing medium. Plants do not communicate, although they do show symptoms of distress. The grower must be able to detect the needs of the plants. Mostly, the grower learns to anticipate the proper growing medium and the nutritional needs of the greenhouse crops.

GROWING MEDIUM

5–2. The growing medium provides plant support, water, nutrients, and gases. (Courtesy, *Greenhouse Product News*)

The health and quality of floriculture crops rest largely with the growing medium. The ***growing medium*** is the material in which plants are grown. In field production and the home garden, soil is the growing medium. In greenhouse production, a variety of materials are used as growing media. The medium used has a direct impact on the quality of the crop grown. Why is the growing medium so important? What function does the growing medium play in plant growth that makes it so important? The importance of the growing medium rests with four basic functions.

- Growing medium holds water for plant use.
- Growing medium provides nutrients for the plant.
- Growing medium permits the exchange of gases to and from the plant roots.
- Growing medium provides support for the plant.

GREENHOUSE GROWING MEDIA ARE DIFFERENT

Have you ever taken soil from outside to use for growing plants in pots? If so, you might have noticed that the soil became hard, that it did not soak water up very quickly, or the plants grew poorly. When used in containers, even good soils lose characteristics important for quality plant growth. Since most greenhouse crops are grown in containers, the growing medium for greenhouse purposes must be different from soils found outside.

Production of greenhouse crops has some specific demands that are often different from the production of outside plants. Rapid growth is required of greenhouse crops. Growers produce a crop quickly, ship it, and bring in a new crop. Therefore, the growing medium must provide conditions that encourage plant growth. Another requirement of greenhouse crops is that they be uniform. A crop in which all the plants have the same rate of growth and development is highly valued by customers. To ensure that the plants grow at the same rate, the growing medium used must be uniform.

DESIRABLE PROPERTIES OF MEDIA

Research has improved greenhouse growing medium. Through research, desirable properties of high quality medium have been identified. In addition, growers have been given a greater understanding of medium and its properties. Discussion of desirable medium properties follows:

Organic Matter

The organic matter in the medium must be stable and must function throughout the life of the crop. ***Organic matter*** is decayed or partially decayed remains of plants and animals found in the medium. It should remain stable and not break down before the crop is finished. Peat moss and bark are the most common sources of organic matter in growing medium.

5–3. The medium shown consists of mostly organic matter.

Bulk Density

Bulk density is the ratio of the mass of dry solids in a medium to the volume of the medium. Light bulk density eases handling and shipping of the potted plants. High or heavy bulk density is needed to provide support for plants. Easter lilies need a medium with a heavy bulk density. The growing medium's bulk density should be light enough to ease handling and heavy enough to support the plant.

5–4. Easter lilies need a medium with a heavy bulk density.

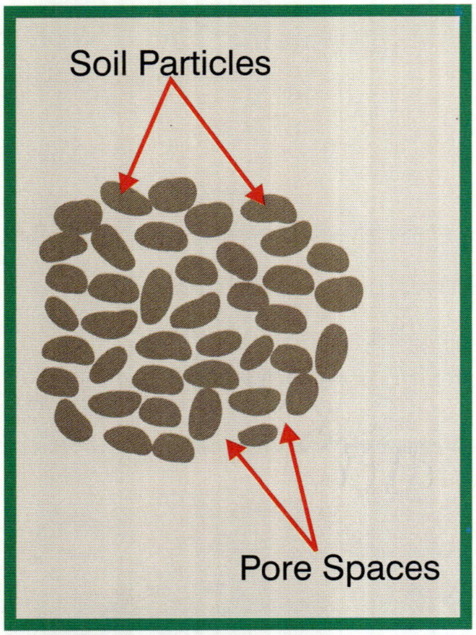

5–5. A higher percentage of pores or porosity in a medium results in better water drainage and aeration or the exchange of gases.

Porosity

The spaces between the solid particles of a growing medium are pores. A higher percentage of pores or ***porosity*** in a medium results in better water drainage and aeration or the exchange of gases. Good mineral or garden soils have about 50 percent pore space. Organic media used in greenhouses have between 75 and 85 percent pore spaces.

Aeration

Aeration is the exchange of gases in the medium. Pore spaces that allow air pockets within the medium are vital for healthy root growth. Root cells use oxygen from the pore spaces to convert sugars to energy. This chemical process is known as respiration. A byproduct of this respiration is carbon dioxide. It is important that a medium has sufficient pore spaces to allow an exchange of carbon dioxide and oxygen.

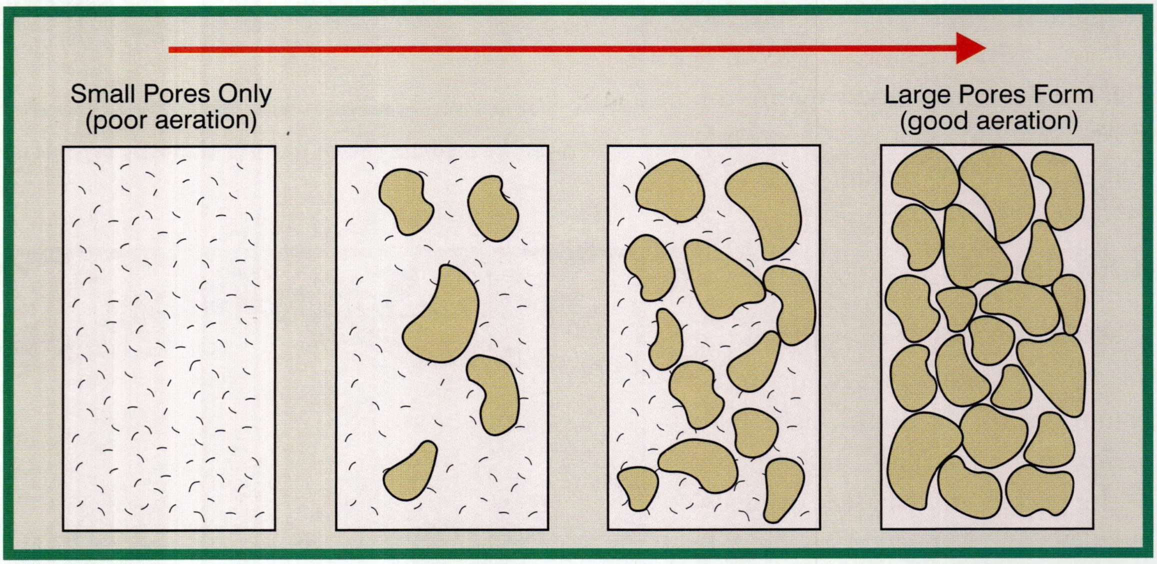

5–6. Pore spaces that allow air pockets within the medium are vital for healthy plant growth.

Water-Holding Ability

A growing medium must store water that is available for plant use. Available water is water that can be absorbed by the plant roots. ***Available water*** is found in the pore spaces of the

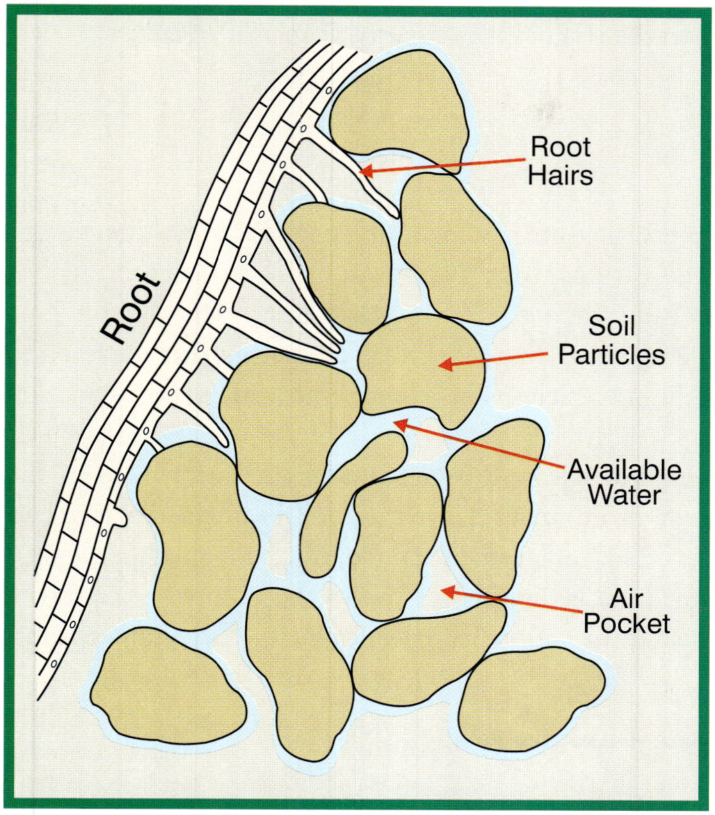

5–7. Available water is water that can be absorbed by the plant roots.

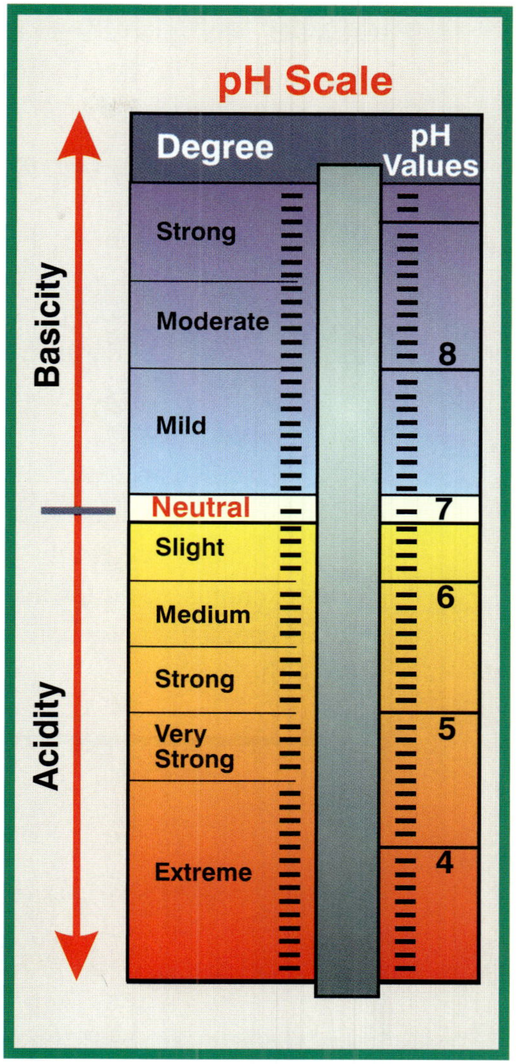

5–8. The pH scale.

medium. Some water in the growing medium is considered unavailable for plant use. ***Unavailable water*** is a thin film of water that binds so tightly on media particles that it cannot be used by plants.

pH

The acidity or alkalinity of the medium is measured as ***pH***. A 14-point scale is used to measure pH. A neutral pH is 7.0. Any reading between zero and 7.0 is acid, and a solution between 7.0 and 14.0 is said to be alkaline or basic. The pH is determined by the concentration of hydrogen (H^+) ions and hydroxyl ions (OH^-) in the soil solution. A sample of pure water has an equal number of H^+ and OH^- ions and is therefore neutral.

Medium pH plays a large role in the availability of nutrients in the medium. Most essential elements for plant growth are available to most plants when the soil pH is between 5.5 and 7.0. If the pH is above or below that range, the availability of nutrients may be restricted. Most plants grow best when the soil pH is between 6.0 and 7.0. Azaleas do best when the pH is 5.5 to 6.0.

Cation Exchange Capacity

Cation exchange capacity is the measure of a medium's capacity to hold nutrients. Many nutrients are positively charged cations, such as potassium (K^+), Calcium (Ca^{+2}), Magnesium (Mg^+), Copper (Cu^+), Iron (Fe^{+2} or Fe^{+3}), Manganese (Mn^{+2}), and Zinc (Zn^{+2}). Particles in the medium have negative charged sites. The cations are attracted to these negatively charged sites on medium particles. This is the same principle of attraction that applies to magnets.

The cations can leave the medium particle and be replaced by a cation held in the medium solution. For instance, a potassium atom may leave the particle and be replaced by a copper atom dissolved in the soil water. This replacement of one cation for another is called cation exchange. The fertility of a medium is directly related to the number of cations a medium can attract and hold. The greater the cation exchange capacity, the greater the fertility of the medium.

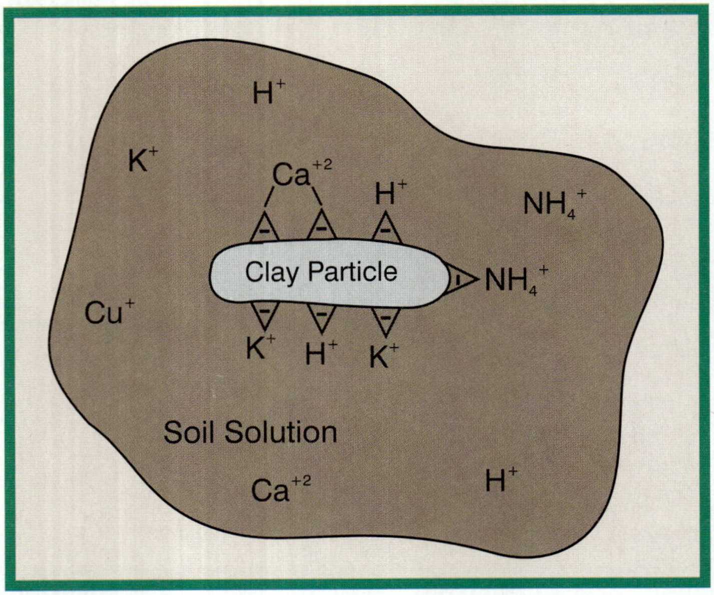

5–9. Negatively charged particles in the growing medium attract and hold positively charged cations.

GROWING MEDIUM COMPONENTS

Most medium used today consists of a mixture of two or more components. More than one component is needed to produce a medium that has desirable properties for plant growth. A common ratio for soil mixes is 1/3 soil, 1/3 peat, and 1/3 perlite or vermiculite.

However, soil is used less frequently now than ever before. Field or mineral soils are seldom uniform, and finding quality soil, free of herbicide residues, has become more difficult. Those media that lack any mineral soil are often referred to as ***soilless mixes***.

Soilless mixes are the standards in the greenhouse industry today. One advantage to using soilless media is they are uniform or always the same. Soilless mixes consist largely of peat moss or bark with perlite and/or vermiculite. The basic components are sterilized to eliminate disease, insect, and weed problems. After sterilizing the materials, nutrients are added and the pH is adjusted for optimal plant growth. Soilless media tend to lack reserves of nutrients for plant use. Typically, treble phosphate and slow release micronutrients are added to media. Also, PNO_3 is often added as a starter fertilizer. For some crops, peat-based soilless media have too low a pH for optimum growth. The pH is raised to acceptable levels by adding limestone or gypsum.

Growers either purchase prepared growing medium in bulk or mix their own medium. Materials commonly used in growing medium mixes include mineral soil, sand, sphagnum peat moss, coir, vermiculite, pine bark, perlite, plastic foam beads, calcined clay, and rock wool. Combined in the right proportions, these materials deliver the desired properties needed for plant growth. The decision of which materials to use is determined by the availability of the material, the cost of the material, shipping costs, and the grower's preference. Materials used have certain characteristics that contribute to a quality medium.

Peat Moss

One of the most widely used components of medium is ***peat moss***. It is an organic material dug from peat bogs. Canada, and to a lesser extent, Michigan and Florida are

5–10. This photo shows peat moss being dug in Canada. (Courtesy, Berger Peat Moss)

5–11. One of the most widely used components of medium is peat moss.

sources of peat for growers in the United States. Canadian peat is very uniform and of high quality. Peat moss has light bulk density, good moisture-holding ability, good air space qualities for the exchange of gases, adequate cation exchange capacity, and a stable pH that is usually between 3.5 and 4.5.

5–12. Coir, made from the husks of coconuts, has similar properties to peat moss. (Courtesy, Crystal Company)

Coir

Coir is made from waste products of the coconut industry and is therefore considered a renewable resource. Its characteristics are similar to those of peat moss. It has high water-holding ability and excellent drainage. Another advantage to coir is that it encourages faster rooting of plants. A disadvantage is it has high salt content, as coconuts grow near the sea. Coir has been used extensively in England and Holland and has gained popularity in the United States.

Vermiculite

Vermiculite is a common ingredient in growing medium. The origin of vermiculite is a mineral called mica. Mica is heated to 1800°F, which causes it to expand like an accordion. Spaces made by the expansion result in good water-holding ability and aeration. It is a lightweight material and can be easily compressed.

Vermiculite is available in different grades, ranging from a fine grade for use with seeds to a coarse grade measuring more than 1/4 inch in diameter. Vermiculite has a good cation exchange capacity and a neutral to slightly alkaline pH (6.3 to 7.8). It also provides small amounts of calcium, magnesium, and potassium for plant use. African vermiculite has a high potassium content.

5–13. Vermiculite is lightweight and has good water holding ability and aeration.

Perlite

Perlite originates from volcanic rock. The rock is crushed, then quickly heated to temperatures of 1800°F. The heat causes it to pop like popcorn to form a lightweight, white aggregate. Perlite is a good lightweight substitute for sand. It is stable, sterile, has little cation exchange capacity, has a pH of 7.5, and provides good water drainage and aeration. Perlite contains fluorides that cause leaf damage to some monocotyledon plants including freesias, dracaenas, and spider plants.

5–14. Perlite is a good lightweight substitute for sand.

Bark

Bark from trees is obtained as a byproduct through the timber industry. It is relatively inexpensive and, thus, lowers the cost of the growing medium. The most widely used bark in the greenhouse industry is pine bark. Bark that is ground into fine pieces is the most useful because of its uniform qualities. Bark improves the moisture-holding ability and aeration of a mix. After peat moss, bark is the second best organic-matter medium component.

5–15. Pieces of bark can be seen in the medium shown.

Sand

Sand is found naturally and is a result of the wearing of rock. Sand is heavy and has a high bulk density that provides solid support for larger plants. It improves water drainage and aeration when used with soil. The sand particles can actually reduce aeration in peat or bark mixes by filling pore spaces created by the peat or bark. The pH of sand is between 7.5 and 8.5. If used, sand must be sterilized to destroy pythium and rhizoctonia disease organisms.

5–16. Sand is heavy and has a high bulk density.

Table 5–1. Characteristics of Medium Components

Material	Bulk Density	Aeration Qualities	Water-Holding Qualities	pH	Cation Exchange Capacity
Bark	Low	High	High	Slightly acid	Low
Calcined Clay	Moderate	High	Low	Varies	Good
Coir	Low	High	High	Slightly acid	Low
Mineral Soil	High	Low	Varies	Varies	Good
Peat Moss	Low	High	High	Acid	Low
Perlite	Low	High	Low	Slightly alkaline	None
Plastic Foam	Low	High	Low	Neutral	None
Rock Wool	Low	High	High	Slightly alkaline	Low
Sand	High	High	Low	Alkaline	Low
Vermiculite	Low	High	High	Slightly alkaline	High

Plastic Foam

Plastic foam flakes or beads (the most common is Styrofoam) are a synthetic polystyrene material. Plastic foam is very lightweight and stable. While it provides good drainage and aeration, it lacks cation exchange capacity, water-holding abilities, and has a neutral pH. It also tends to float to the surface of growing media.

Calcined Clay

Calcined clay is clay aggregate heated to form a hard, stable particle. It has a high bulk density. Calcined clay improves water drainage and aeration in a medium. It also has some cation exchange capacity.

Rock Wool

Rock wool is a human-made material from an igneous rock, basalt. The rock is heated to temperatures of 2700°F and, once liquefied, is spun into fibers similar in appearance to cotton candy. Rock wool is used extensively in hydroponic operations. Rock wool cubes are used in plant propagation. Granular forms of rock wool are used as a medium component. Rock wool has good water-holding capacity and good aeration. It is slightly alkaline and has a low cation exchange capacity.

Mineral Soil

Mineral soil is obtained from nature and is the result of weathered rock. It has become more difficult for growers to find uniform mineral soil with a favorable pH and good structure and texture. Mineral soil must be free of agricultural herbicides and sterilized before use.

5–17. Mineral soil must be free of agricultural herbicides and sterilized before use.

Soil Sterilization

Mineral soil and sand must be sterilized before use. Soil sterilization is a term used to describe the killing of living things in soil. The idea is to kill bacteria, fungi, insects, nematodes, and weed seeds that might be harmful to the crops. Steam sterilization is the most common method of sterilizing soil. Steaming soils to a temperature of 180°F for 30 minutes sufficiently kills most soil organisms. The increased use of soilless mixes has greatly reduced the need for sterilizing soils for pot plants and bedding plants.

PLANT NUTRITION

One aspect of plant production over which growers can have great control is plant nutrition. ***Plant nutrition*** involves the absorption of nutrients for plant growth. The nutritional needs of greenhouse plants require close attention because the plants are wholly dependent on what is available in the growing medium. Current fertilizer practices allow growers to adjust and deliver nutrients to maintain desired plant growth.

Plant growth is dependent on 16 essential elements, often referred to as nutrients. Three of the elements, oxygen (O), hydrogen (H), and carbon (C), make up 89 percent of a plant's tissue by dry weight. They are considered to be nonfertilizer nutrients. The plant acquires these three elements through natural processes from air and water.

5–18. CO_2 burners release carbon dioxide gas into the greenhouse atmosphere.

Nine of the 16 essential elements are required in greater quantity than the others. They are called ***macronutrients***. Oxygen, hydrogen, and carbon are three. The other macronutrients are nitrogen (N), phosphorus (P), potassium (K), calcium (Ca), magnesium (Mg), and sulfur (S). Nitrogen, phosphorus, and potassium are considered to be ***primary macronutrients*** because they are used in complete fertilizers. Calcium, magnesium, and sulfur are said to be secondary macronutrients because plants need moderate amounts. These ***secondary macronutrients*** may or may not be used in complete fertilizers.

Table 5–2. Essential Elements

	Element	Chemical Symbol	Percent of Plant's Dry Weight
Nonfertilizer	Carbon	C	89
	Hydrogen	H	
	Oxygen	O	
Macronutrients	Nitrogen	N	4.0
	Phosphorus	P	0.5
	Potassium	K	4.0
	Calcium	Ca	1.0
	Magnesium	Mg	0.5
	Sulfur	S	0.5
Micronutrients	Iron	Fe	0.02
	Manganese	Mn	0.02
	Chlorine	Cl	0.01
	Boron	B	0.006
	Zinc	Zn	0.003
	Copper	Cu	0.001
	Molybdenum	Mo	0.0002

The other seven essential elements are needed in small quantities. They are called ***micronutrients***. Sometimes they are called trace elements. They are boron (B), copper (Cu), chlorine (Cl), iron (Fe), manganese (Mn), molybdenum (Mo), sodium (Na), and zinc (Zn). Six of these, boron, copper, iron, manganese, molybdenum, sodium, and zinc, are supplied to plants as fertilizers. Chlorine need not be added as a fertilizer. Plants obtain sufficient quantities of chlorine from the medium or water.

A little phrase can be used to help memorize the 16 essential elements for plant growth. It is "C. B. Hopkins Café Mighty Good Closed Monday Morning See You Zen." It represents the following: Carbon (**C**), Boron (**B**), Hydrogen (**H**opkins), Oxygen (H**O**pkins), Phosphorus (Ho**P**kins), Potassium (Hop**K**ins), Nitrogen (Hopki**N**s), Sulfur (Hopkin**S**), Calcium (**Ca**fé), Iron (caf**é**), Magnesium (**M**ighty **g**ood), Chlorine (**Cl**osed), Manganese (**Mon**day), Molybdenum (**Mo**rning), Copper (See you = **Cu**), Zinc (**Zen**).

Other elements are not listed as essential but play important roles in plant growth and development. For example, recent studies have shown silicon (Si) to improve plant strength and disease resistance. Nickel (Ni) is another element considered important for plant growth.

FERTILIZERS

A ***fertilizer*** is any material added to growing medium that provides nutrients for plants. Adding nutrients in the form of fertilizers is necessary to improve plant growth and development. Fertilizers are particularly important with the use of soilless mixes. Because soilless mixes lack reserves of nutrients, the nutrients must be supplied regularly to the plants. Also, quick production of quality crops is achieved when all the nutrients required of plant growth are provided at the optimum rates.

5–19. Water soluble fertilizers used in the greenhouse often come with a blue indicator dye. (Courtesy, University of Illinois Extension)

Table 5–3. Common Fertilizers Used in Greenhouse Crop Production

	Percent Nitrogen	Percent Phosphorus	Percent Potassium	Other Characteristics
Water-Soluble Fertilizer Compounds				
Ammonium nitrate	33	0	0	Acid
Ammonium sulfate	21	0	0	Acid
Calcium nitrate	15	0	0	Basic Provides Calcium
Diammonium sulfate	21	53	0	Acid
Magnesium nitrate	11	0	0	Neutral
Monoammonium phosphate	12	62	0	Acid
Monopotassium phosphate	0	53	35	Basic
Potassium chloride	0	0	60	Neutral
Potassium nitrate (saltpeter)	13	0	44	Basic
Potassium sulfate	0	0	51	Neutral
Sodium nitrate	16	0	0	Basic
Urea	45	0	0	Acid

(Continued)

Table 5–3 (Continued)

	Percent Nitrogen	Percent Phosphorus	Percent Potassium	Other Characteristics
Slow Release Fertilizers				
Osmocote (coated)	14	14	14	3–4 month release
Sierra (coated)	17	6	10	Plus micronutrients 8–9 month release
Nutricote (coated)	18	6	8	Plus micronutrients 100 day release
ProKote	20	3	10	Plus micronutrients 7–9 month release
MagAmp	7	40	6	3–5 month release

Different plants have different nutrient level requirements. Also, plants have different nutrient requirements based on their stage of growth. For potted flowering plants, there are five stages of growth. In the first or juvenile stage, the plant is a young seedling or cutting. In the second stage, vegetative or leafy growth is encouraged. The vegetative stage is followed by the flower bud initiation stage, the flower bud development stage, and the flowering stages. Each stage calls for different rates of fertilizers. When in the vegetative stage, plants use more nitrogen, and when in the flowering stage, they require less nitrogen and more phosphorus.

There are two widely used methods to deliver fertilizers to greenhouse plants. One involves the use of water-soluble fertilizers with an injector system, whereby the plants are fertilized as they are watered. The other involves the use of slow release fertilizers.

Fertilizer Injector Systems

Fertilizer injector systems are standard equipment in today's greenhouses. There are many types of injectors on the market. Basically, they work the same. A concentrate of water-soluble fertilizer is mixed in a tank. ***Water-soluble fertilizers*** dissolve completely in water and stay in solution. The concentrations of the water-soluble fertilizers are also easily adjusted. The fertilizer concentrate is

5–20. Fertilizer injectors deliver exact amounts of nutrients to plants.

then mixed with the water in exact proportions before irrigating plants. With fertilizer injectors, growers can provide exact levels of nutrients.

There are a number of advantages to using fertilizer injectors. The greenhouse plants are provided with constant liquid feeding. This is especially important with soilless media that have a low cation exchange capacity. The plants are fertilized when they are watered thus reducing labor. All the plants of a particular crop receive the same levels of nutrients. This assists in producing a uniform crop. In addition, adjustments to the level of nutrients in solution can be made easily.

Calculating ppm

Nutrients in solution are measured in ***parts per million*** or ppm. Growers can refer to tables in reference books or to instructions with their injector system to determine the amount of fertilizer needed to reach a desired ppm. However, growers must also understand how to calculate parts per million (ppm) of fertilizer nutrients.

An example problem might call for 200 ppm nitrogen. The chosen fertilizer is calcium nitrate with an analysis of 15-0-0. A rule of thumb in calculating ppm is that *1 oz. of anything in 100 gallons H_2O equals 75 ppm.*

Problem: Wanted, 200 ppm N
15-0-0 Calcium Nitrate

Multiply the percent of the nutrient in the given fertilizer by 75.
.15 N × 75 = 11.63 ppm N (if 1 oz. added to 100 gallons)
200 ÷ 11.63 = 17.2 oz. of fertilizer needed in 100 gallons to get 200 ppm

If the injector system delivers 1:100 (one gallon concentrate mixed with 99 gallons water for a total of 100 gallons), 17.2 ounces of fertilizer is needed for each gallon of concentrate. If the tank holds 30 gallons of concentrate, 516 ounces (32.25 pounds) of fertilizer are needed to deliver 200 ppm.

Slow Release Fertilizers

Slow release fertilizers are self-defined by their name. These products continually discharge a small amount of nutrients over a period of time. They are available in two forms. One form consists of a water-soluble fertilizer held within a plastic resin or sulfur coating. The coating is designed to allow a small amount of fertilizer to be slowly released. They have the appearance of little round beads and are sometimes mistaken as insect eggs by the uninformed. The other type of slow release fertilizer comes in a granular form. This type releases no more than 1 percent of the fertilizer in a 24-hour period.

The use of slow release fertilizers has declined with the increased use of water-soluble fertilizers through injector systems. Growers have much greater control with fertilizer injector systems. However, there are some advantages to slow release fertilizers. Slow release fertilizers can be safely mixed with the growing medium prior to planting. Slow release fertilizers provide a more constant supply of nutrients than water-soluble fertilizers. This is particularly true when watering is less frequent. Slow release fertilizers work well when capillary mats are used to water plants. Also, slow release fertilizers can be used to extend the shelf life of plants once they leave the greenhouse.

5–21. Slow release fertilizers can be mixed with the growing medium prior to planting or applied to the medium surface after potting.

NUTRIENT DEFICIENCY

Plant quality and thus, profits, are lost if the plants cannot obtain the essential nutrients from the medium. Even if only one necessary nutrient is missing, the plant can develop deficiency symptoms. For example, nitrogen promotes the growth of green, leafy tissue. If the available nitrogen is low, the plant may take on a yellow appearance. The growth rate slows and the plant often is stunted. This happens even when all the other essential nutrients are available.

If a plant fails to receive the needed nutrients, it will show signs of ***nutrient deficiency***. Nutrient deficiencies most often surface as an unhealthy plant appearance. Symptoms vary with the nutrient that is in short supply. Common symptoms of deficiencies include discoloration of the leaves, death of leaf tissue, and stunted growth. Because of the complex interactions of nutrients in plant processes, deficiency symptoms for different nutrients are often very similar. Laboratory tests, including the analysis of plant tissues, can be used to determine which nutrient is lacking.

5–22. Discoloration of leaves is often a result of nutrient deficiency.

Table 5–4. Nutrient Influences/Functions in Plants and Symptoms of Deficiency

Nutrient	Influence/Function	Deficiency Symptom
Primary Macronutrients		
Nitrogen	Produces stem and leaf growth; gives plants dark green color; synthesis of amino acids and proteins	Entire plant lighter green, lower leaves yellowing; slow or dwarfed growth
Phosphorus	Stimulates root development and growth; aids in cell division; encourages flower bud formation; improves winter hardiness; helps plants to a vigorous and rapid start	Purplish coloration to leaves and stems; stunted growth
Potassium	Increases plant vigor and disease resistance; aids in the transport of foods through the phloem; has key role in opening and closing stomata; thickens cell walls	Yellowing or death of tissues at tips and outer edges of older leaves
Secondary Macronutrients		
Calcium	Maintains strength of cell walls; promotes early root growth	Short, much branched roots; young leaves at growing points die back
Magnesium	Essential for chlorophyll and photosynthesis; activator for many plant enzymes; aids nutrient uptake	Loss of green leaf color starting with bottom leaves
Sulfur	Stimulates root growth; needed for protein formation; gives green color	Young leaves light green with veins being lighter
Micronutrients		
Boron	Essential for pollination and reproduction; cell division; transport of sugars	Young leaves yellow and thick
Copper	Regulates several chemical processes; chlorophyll synthesis and respiration	Yellowing of leaves,, younger leaves affected first
Chlorine	Involved in light reactions of photosynthesis; aids root and shoot growth	Symptoms have not been recognized
Iron	Important in chlorophyll formation; component of enzymes involved in photosynthesis, respiration and nitrogen fixation	Young leaves yellow first; veins remain green
Manganese	Important in chlorophyll formation; part of enzymes involved in respiration and nitrogen metabolism	Young leaves yellow first; veins remain green
Molybdenum	Part of enzymes involved in nitrogen metabolism; aids nitrogen fixation and protein synthesis	Older leaves yellow; stunted growth
Zinc	Important in chlorophyll, auxin, and starch formation; part of enzymes involved in respiration; needed for auxin and starch formation	Older leaves yellow; stunted growth

SOLUBLE SALTS

A problem that arises with the use of water-soluble fertilizers is the build-up of soluble salts. The nutrients that are in solution along with mineral salts from the water and medium contribute to the soluble salt content in the growing medium. ***Soluble salts*** consist of these dissolved mineral salts. High levels of soluble salts are harmful to plants. Young plants are especially sensitive.

The damage caused by soluble salts is through a process known as reverse osmosis. As the salt concentration builds up in the growing medium, it prevents water molecules from being absorbed by the roots. In severe cases, water is drawn from the root cells into the growing medium. The resulting damage caused by soluble salts is the burning of the roots. Wilting or the death of leaf tissues follows the injury or death of the roots.

Damage from soluble salts can be avoided by preventing a buildup of soluble salts in the growing medium. A good practice is to water the plants until water drains from the container. As the water flows through the growing medium, it carries soluble salts with it. This is known as leaching.

5–23. Watering heavily so that water drains through the bottom of the pot reduces soluble salt build up.

REVIEWING

MAIN IDEAS

Growing medium serves four major functions for plants. It holds water for plant use and provides nutrients for the plant. Growing medium permits the exchange of gases to and from the plant roots. Also, growing medium provides support for the plant. Some characteristics of growing medium are desirable. Properties to consider include organic matter, bulk density, porosity, aeration, water-holding ability, pH, and cation exchange capacity.

Greenhouse growing media consist of different components. Materials commonly used in growing media mixes include mineral soil, sand, sphagnum peat moss, coir, vermiculite, pine bark, perlite, plastic foam beads, calcined clay, and rock wool. Combined in the right proportions, these materials deliver the desired properties needed for plant growth.

Plant nutrition involves the absorption of nutrients for plant growth. Greenhouse plants are wholly dependent on what is available in the growing medium for their nutritional needs. Plant growth is dependent on 16 essential elements, referred to as nutrients. Three of the elements, oxygen, hydrogen, and carbon make up 89 percent of a plant's tissue by dry weight. Nine of the 16 essential elements are called macronutrients because they are used in greater quantity. In addition to oxygen, hydrogen, and carbon, the macronutrients are nitrogen (N), phosphorus (P), potassium (K), calcium (Ca), magnesium (Mg), and sulfur (S). The other seven essential elements are needed in small quantities and are called micronutrients. They are boron (B), copper (Cu), chlorine (Cl), iron (Fe), manganese (Mn), molybdenum (Mo), sodium (Na), and zinc (Zn).

Fertilizers are materials added to growing medium that provide nutrients for plants. Water-soluble fertilizers dissolve completely in water and stay in solution. Fertilizer injector systems are standard equipment in today's greenhouses for providing exact levels of water-soluble nutrients. Nutrients in solution are measured in parts per million or ppm. Slow release fertilizers continually discharge a small amount of nutrients over a period of time. Nutrient deficiencies and soluble salts are two problems associated with fertilization programs.

QUESTIONS

Answer the following questions using complete sentences and correct spelling.

1. What are the functions of a growing medium?
2. What are the desirable properties of a growing medium?
3. Why is aeration in a growing medium important?
4. Why is pH of a growing medium important?
5. Why are soilless mixes a standard in the greenhouse industry today?
6. How do macronutrients and micronutrients differ?
7. What advantage is there to using water-soluble fertilizers?
8. How do slow release fertilizers work?
9. What are symptoms of nutrient deficiency?
10. Why are high concentrations of soluble salts considered to be bad?

EVALUATING

Match the term with the correct definition. Write the letter of the term on the line provided.

a. aeration	e. growing medium	i. bulk density
b. cation exchange capacity	f. vermiculite	j. organic matter
c. coir	g. soluble salts	k. parts per million
d. micronutrients	h. fertilizer	l. slow release fertilizers

______1. Made from waste products of the coconut industry and is similar in properties to peat moss.

______2. A common ingredient in growing media that originates as a mineral called mica.

______3. Any material added to growing media that provides nutrients for plants.

______4. The measure of a medium's capacity to hold nutrients.

______5. Eight essential elements, sometimes they are called trace elements, needed in small quantities.

______6. The ratio of the mass of dry solids in a medium to the volume of the medium.

______7. The material in which plants are grown.

______8. The measure of nutrients in solution.

______9. Consist of dissolved mineral salts.

_____10. Fertilizers that continually discharge a small amount of nutrients over a period of time.

_____11. The exchange of gases in the medium.

_____12. Decayed or partially decayed remains of plants and animals found in the medium.

EXPLORING

1. Obtain samples of different growing medium components by contacting manufacturers or buying materials at a retail store. Touch and look at the materials. Weigh the materials. Soak the materials with water. Note your observations based on the desirable properties of a growing medium discussed in this chapter. You might also try growing plants in the different components.

2. Set up an experiment in the classroom or at home whereby four plants of the same species are given different fertilizer rates. Start with plants of the same size, in the same type of medium, and in identical containers. Calculate the ppm of four different fertilizer solutions (gallon milk jugs work well). For example, the plants might be given no fertilizer, 200 ppm nitrogen, 400 ppm nitrogen, and 800 ppm nitrogen. Grow the plants in the same location and water with the nutrient solution as needed. Observe the plants closely and record your findings.

CHAPTER 6

Plant Anatomy

OBJECTIVES

This chapter is about plant anatomy. It has the following objectives:

1. Identify the major plant organs and their components.
2. Describe the functions of the root system.
3. Discuss the functions of the stem.
4. Describe the functions of the leaves.
5. Identify different types of leaves.
6. Describe the functions of the flower.
7. Identify different types of flowers.
8. Relate plant functions to floriculture practices.

TERMS

adventitious roots
anther
anthocyanin
apical dominance
apical meristem
auxin
basal rooters
bract
bud
bud scale
bulbs
calyx
cambium
carotene
catkin
chlorophyll
complete flowers
compound leaves
corms
corolla
corymb
cuticle
cyme
determinate
dicot
epidermis
filament
flavonol
flowers
guard cells
head
imperfect flower
incomplete flower
indeterminate
inflorescence
lateral bud
leaf blade
leaves
margin
mesophyll
monocot
ovary
ovule
palisade layer
palmately compound
panicle
pedicel
peduncle
perianth
perfect flowers
petals
petiole
phloem
photosynthesis
pinnately compound
pistil
pollen
primary root
raceme
receptacle
rhizome
root cap
root hairs
root system
root tuber
secondary root
sepals
simple leaves
solitary flower
spadix
spathe
spike
spongy layer
stamens
stem tubers
stigma
stolon
stomata
style
tepal
terminal bud
transpiration
umbel
variegation
xanthophyll
xylem

6–1. People in the floriculture industry understand the functions of plant parts. (Courtesy, Cargill)

ARE YOU interested in floriculture? If you are, you will need to know some of the basic information known by those in the industry. People in the industry deal with plants. They know the parts of a plant. They understand the functions of the parts. This knowledge allows them to do two things. They can communicate clearly with other people in the floriculture industry. It also improves their ability to grow, handle, and work with floriculture crops.

The study of plant anatomy also helps to satisfy a curiosity to know what makes plants tick. The knowledge helps to answer questions. Why does a plant have roots? How does a plant move materials from its leaves to its roots? Why are leaves green? How are flowers involved in seed production? The answers to these and many other questions are more easily understood with the study of plant anatomy.

PLANT ORGANS AND THEIR FUNCTIONS

Like people, plants are complex organisms. Plants are made of organs consisting of tissues and cells. People have organs consisting of tissues and cells, too. Our heart, liver, kidneys, and other organs carry out important functions enabling us to live. If one organ fails, our life is at risk. Plant organs, including roots, stems, leaves, and flowers, are equally important to the plant's survival.

Understanding plant growth and the function of plant organs is very important to the floriculture industry. Growers strive to promote rapid growth and high quality crops. To achieve this goal, they manipulate greenhouse conditions. They also apply their knowledge of plants and the functions of the different plant organs.

6–2. The root system supports above ground growth of a plant. Shown is a Zebra Plant, *Aphelandra squarrosa.*

ROOTS

Very often, the health of a plant can be linked to its root system. The roots can be considered the foundation for plant growth. When discussed collectively, the roots are referred to as a ***root system***. The root system performs important functions for the plant. Water and minerals are absorbed into the plant root system from the growing medium. The root system anchors the plant so it can grow upright. The root system also stores food manufactured in the leaves.

If you were to take a close look at a root, you would notice certain features. The first thing you might notice is that most healthy roots are white or cream in color. Another feature is the abundance of tiny ***root hairs*** found near the growing tip of the root. Root hairs greatly increase the surface area of the root allowing more water and minerals to be absorbed into the plant. Actually, the majority of water and nutrient absorption occurs through the root hairs.

6–3. This photo shows root hairs along a young root. (Courtesy, Agricultural Research Service, USDA)

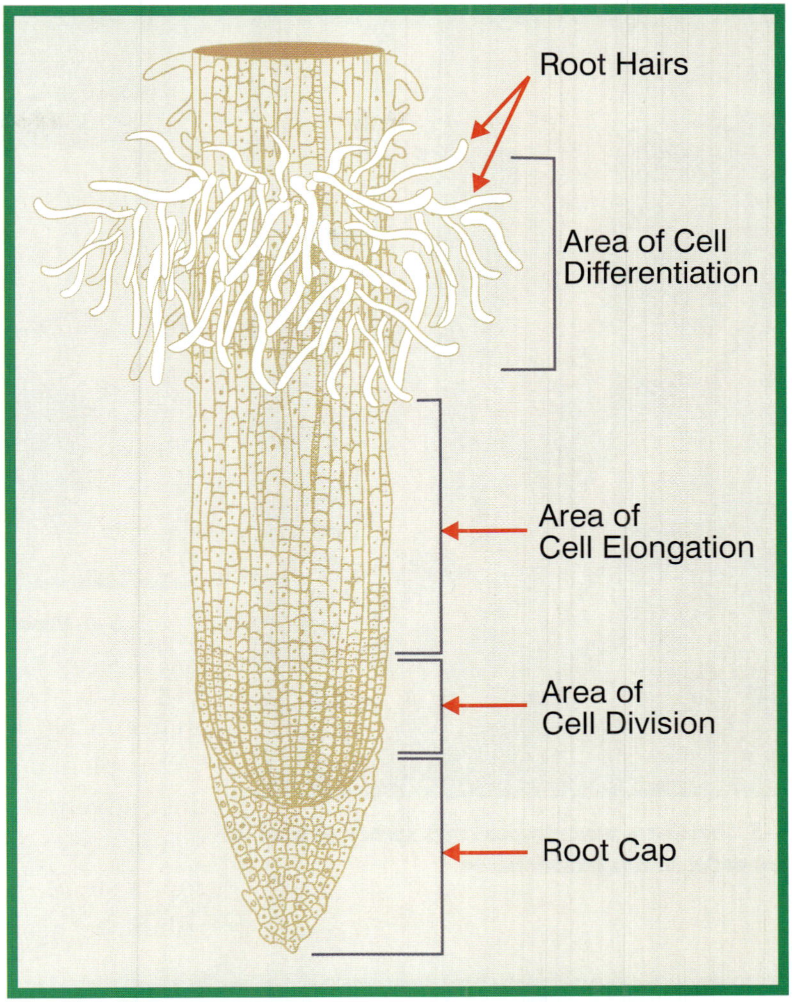

6–4. A diagram of the growing point of a root.

When a seed begins to grow, it sends out a single, main root, called the ***primary root***. The primary root absorbs water and nutrients and anchors the seedling. As the primary root grows, it produces many smaller root branches called ***secondary roots***. It is on the young, developing roots of a plant that root hairs can be seen. As roots grow, their tips are protected from coarse soil. This is accomplished by a mass of cells called the ***root cap***. As the plant grows and the root system develops, the roots grow in all directions forming a tangled mass.

Not all roots begin growth from root tissue. Some roots begin growth from the stem or a leaf. These roots are called ***adventitious roots***. Many greenhouse crops are started or propagated by cuttings. In these cases, a stem or leaf is removed from a mother plant and placed in a growing medium or a specially designed rooting cube. Given the right conditions, adventitious roots grow from the stem or leaf. Cuttings from some plants, such as poinsettias, chrysanthemums, and zonal geraniums, produce adventitious roots at the base of the

6–5. Shown are adventitious roots sprouting from the nodes of this philodendron.

6–6. Basal rooting is characteristic of chrysanthemum cuttings.

stem cutting. These plants are called ***basal rooters***. Other plants, particularly tropical foliage plants, produce adventitious roots along the stem.

Growers recognize the importance of a healthy root system. As a result, they use practices that contribute to healthy root growth. They select growing media and containers best suited for the plant. They plant the plants at appropriate depths. They are also careful to provide the right amounts of water when needed. The importance of the medium and watering practices must not be overlooked. Growing medium kept too wet limits oxygen available to the roots. Without oxygen, the root cells die. Wet conditions are also ideal for plants to become infected with soil-borne diseases, such as *Rhizoctonia* and

6–7. The brown roots on this rooted zonal geranium cutting indicate the presence of disease possibly caused by cold, wet conditions.

Pythium. Medium that dries excessively often results in wilting and death to cell tissues from loss of moisture.

Care is taken to avoid a buildup of soluble salts in the growing medium. Soluble salts are dissolved minerals and fertilizers in the water. Over time soluble salts can build up in the medium. High concentrations of salts have a burning effect on roots. To prevent the buildup of soluble salts, it is recommended to leach or wash the soluble salts from the growing medium by watering thoroughly.

Roots can become large food-storage organs. A structure called a **root tuber** is an enlarged food-storage root with adventitious shoots. As the adventitious shoots develop, they produce adventitious roots that expand to become tubers. Root tubers are used in propagation. Examples of plants with root tubers include dahlia, tuberous begonia, and sweet potato.

STEMS

As with roots, stems serve important functions for the plant. Stems hold the leaves in a position to take the best advantage of sunlight. Stems conduct water and minerals absorbed by the roots to the leaves. Stems conduct food made in the leaves to the rest of the plant. Stems store food and water in their tissues. Stems produce new living tissues.

6–8. Stems hold the leaves and flowers in an upright position.

Conductive Tissues

The life flow of a plant is found in its stems. Just as we need to continually pump blood through our bodies, a plant needs to conduct water, minerals, and food to all of its parts. The plant can accomplish this with special conductive tissues called xylem and phloem. Xylem and phloem tissues in the root are continuous with the xylem and phloem tissues of the stem, which are continuous with the xylem and phloem in the leaves and flowers.

The conductive tissue in the stem that transports water and minerals from the roots to the leaves is called the **xylem**. Xylem cells also have stiff walls that provide structural support for the plant. Xylem is located in the roots, stem, leaves, and flowers. Cut flowers in a vase continue to flower because of the flow of water up through the xylem tissues.

6–9. Conductive tissues in stems keep leaves and flowers supplied with water and minerals.

Food made in the leaves is transported to the rest of the plant through the ***phloem*** tissue. Phloem, like the xylem, is found throughout the plant. The roots would die if the plant lacked the ability to move food to the root system.

The location of xylem and phloem within a stem differs with the type of plant. Monocotyledons or monocot plants have xylem and phloem scattered throughout the stem. ***Monocot*** plants include narrow leaf plants. Easter lilies, iris, spider plants, and dracaenas are examples of monocotyledons. Dicotyledon plants or ***dicot*** plants have xylem and phloem in a ring within the stem. Dicots also have a layer of cells called the ***cambium*** where cell division takes place. These dividing cells become either xylem or phloem cells depending on which side of the cambium they are located. Examples of dicots are hydrangea, florist azalea, African violet, and chrysanthemum.

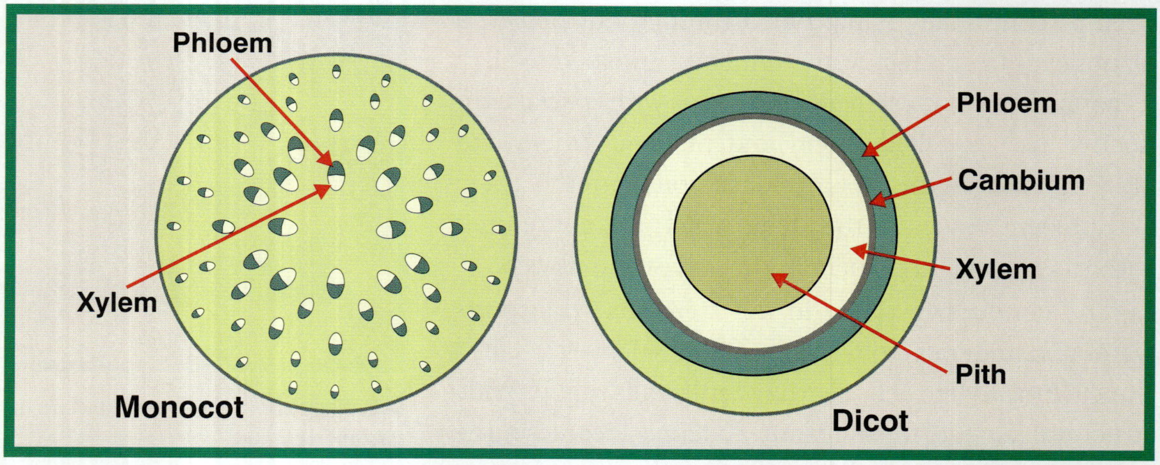

6–10. The location of xylem and phloem within a stem differs between monocots and dicots.

Buds

Stems have structures called ***buds***. Buds contain undeveloped leaves, stem, or flowers. ***Bud scales*** cover and protect these undeveloped parts during dormancy. On some plants, the buds can be difficult to see. Buds tend to produce either vegetative or flowering parts of the plant. Vegetative buds contain immature leaves. Flower buds hold immature flowers. Usually, flower buds are visibly larger than vegetative buds. This is the case with the florist azalea and hydrangea.

The large bud at the tip of a stem is referred to as the ***terminal bud***. The terminal bud is important because it contains the ***apical meristem*** or the primary growing point of the stem. Hormones, called ***auxins***, are produced in the apical meristem. The auxins migrate down the stem and influence the growth of the stem.

Buds located along the sides of a stem are ***lateral buds***. Lateral buds can be found where the leaf petiole is attached to the stem. Lateral buds closest to the apical meristem are inhibited from developing by the concentration of auxins. The apical bud exerts its dominance over the lateral buds. The term to describe this is ***apical dominance***. Further down the stem, lateral buds develop as the influence of the auxins becomes weaker. The lateral buds also will develop if the terminal bud is removed, as is the case when plants are pinched.

6–11. Buds protect undeveloped flowers.

6–12. Auxins, produced in the apical meristem, inhibit the development of lateral buds.

Specialized Stems

Stems can become specialized as in the case of bulbs, corms, rhizomes, and stem tubers. These structures serve as underground food and water storage organs. ***Bulbs*** are short, flattened stems that bear fleshy food storage leaves. At the base of each fleshy leaf there is a bud. Examples of bulbs include the tulip and Narcissus. Corms, although they look like bulbs, differ in structure. A ***corm*** is a short, swollen, underground stem. Crocuses and gladioli are corms. A ***rhizome*** is an underground horizontal stem. Iris, alstroemeria, and calla lilies are rhizomes. ***Stem tubers*** are swollen tips of a rhizome. The Irish potato is a stem tuber.

A stem structure similar to a rhizome is the stolon or runner. The difference is that the ***stolon*** is a stem that grows horizontal above the ground and may produce roots at its tip or at nodes. Airplane, or spider, plants and strawberry begonia are two foliage plants that produce stolons.

LEAVES

Leaves are plant organs responsible for the production of food (sugars) for the plant. All life on Earth is dependent on this food-producing reaction that takes place in the leaves. It is called ***photosynthesis***. Growth and development of greenhouse crops is determined by the rate of photosynthesis.

Leaves are thin organs containing specialized cells. There is an upper and lower ***epidermis*** or a protective layer of cells. The epidermis cells have an exterior coating that is waxy. The waxy coating, called the ***cuticle***, serves to prevent excessive water loss from the leaf tissues. Cacti, jade plants, and kalanchoes have thick cuticles. Other plants have hairy leaves, in addition to the cuticle, that help to reduce water loss. Examples of hairy leaf plants are African violets, gloxinia, zonal geranium, and purple passion plant.

6–13. The purple hairs are easy to see on the leaves of this Purple Passion Plant, *Gynura purantiaca*.

Leaves have the capability to "breathe" too. There are pores or openings in the epidermis called ***stomata*** (singular: stoma). These openings allow the exchange of oxygen, carbon dioxide, and water vapor. Most stomata can be found on the underside of leaves. The movement of water vapor through stomata is referred to as ***transpiration***. The stomata are opened and closed by a pair of ***guard cells***. When water is plentiful and light is shining, the guard cells are pumped full of water. This causes the cells to push apart, creating an opening to allow an exchange of gases.

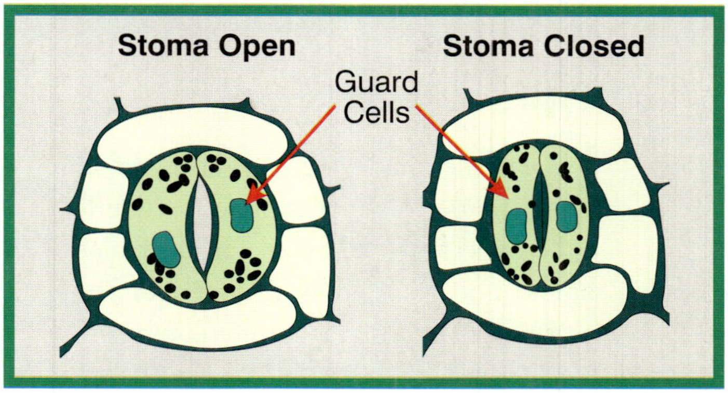

6–14. The stomata are opened and closed by a pair of guard cells.

The stomata close when water leaves the guard cells, causing the cells to collapse. Stomata close at night. They also close when the plant is experiencing water stress. If the medium dries out or if the roots cannot replace the water lost by transpiration, the plant tries to conserve water. The plant might even wilt or become limp. When the stomata close, CO_2 is no longer available, so photosynthesis stops. Wilting can be a serious problem with some greenhouse crops. For instance, if potted chrysanthemums wilt, they cannot reach their full potential.

Sandwiched between the epidermal layers is the photosynthetic tissue of the leaf known as the ***mesophyll***. Mesophyll is a Greek word meaning "the middle of the leaf." The bulk of photosynthetic activity for a plant takes place in the mesophyll cells. Just below the upper epidermis is a stacked layer of mesophyll cells called the ***palisade layer***. Underneath the palisade layer is a loosely arranged layer of mesophyll cells called the ***spongy layer***.

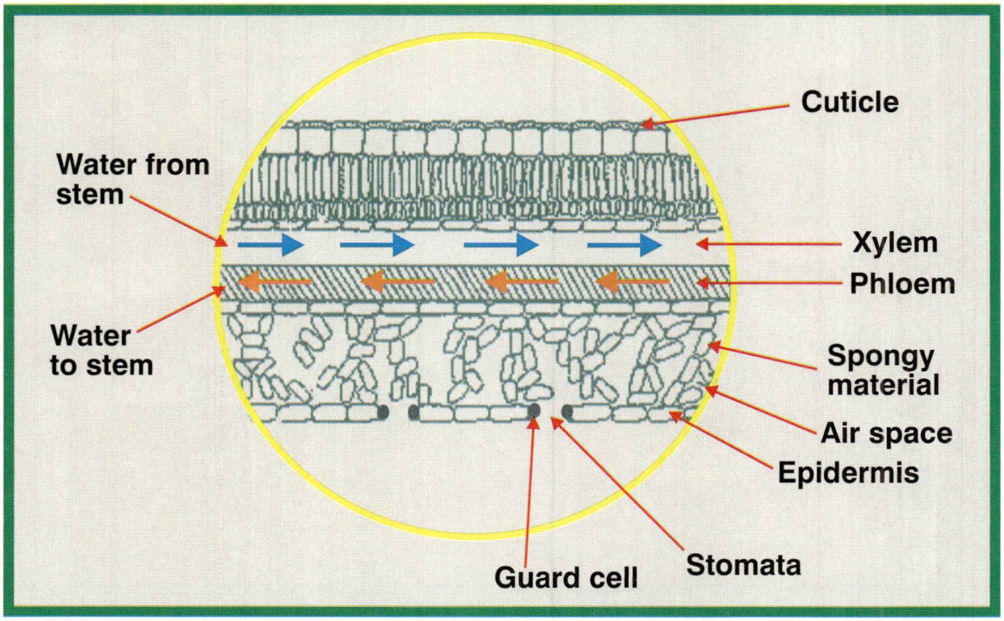

6–15. The bulk of photosynthetic activity for a plant takes place in the mesophyll cells.

Throughout the mesophyll, there is a network of veins. Each vein contains xylem and phloem tissues. The veins are numerous enough so every cell is reached for exchange of materials. The xylem keeps the mesophyll cells supplied with water and minerals for the production of food. The food is then transported to the rest of the plant through the phloem.

Types of Leaves

Leaves come in a great variety of shapes, sizes, and colors. Leaves are also useful when trying to identify plants. There are different parts to a leaf. The large broad part of a leaf is called the ***leaf blade***. The leaf blade provides a large surface area that increases the amount of solar energy absorbed for photosynthesis. The edge of the leaf blade is referred to as the ***margin***. The margins can be one of many forms including wavy, toothed, lobed, and entire or smooth. The leaf blade is connected to the stem by the ***petiole*** or leaf stalk. Water and minerals flow through the xylem in the petiole to the cells in the leaf blade.

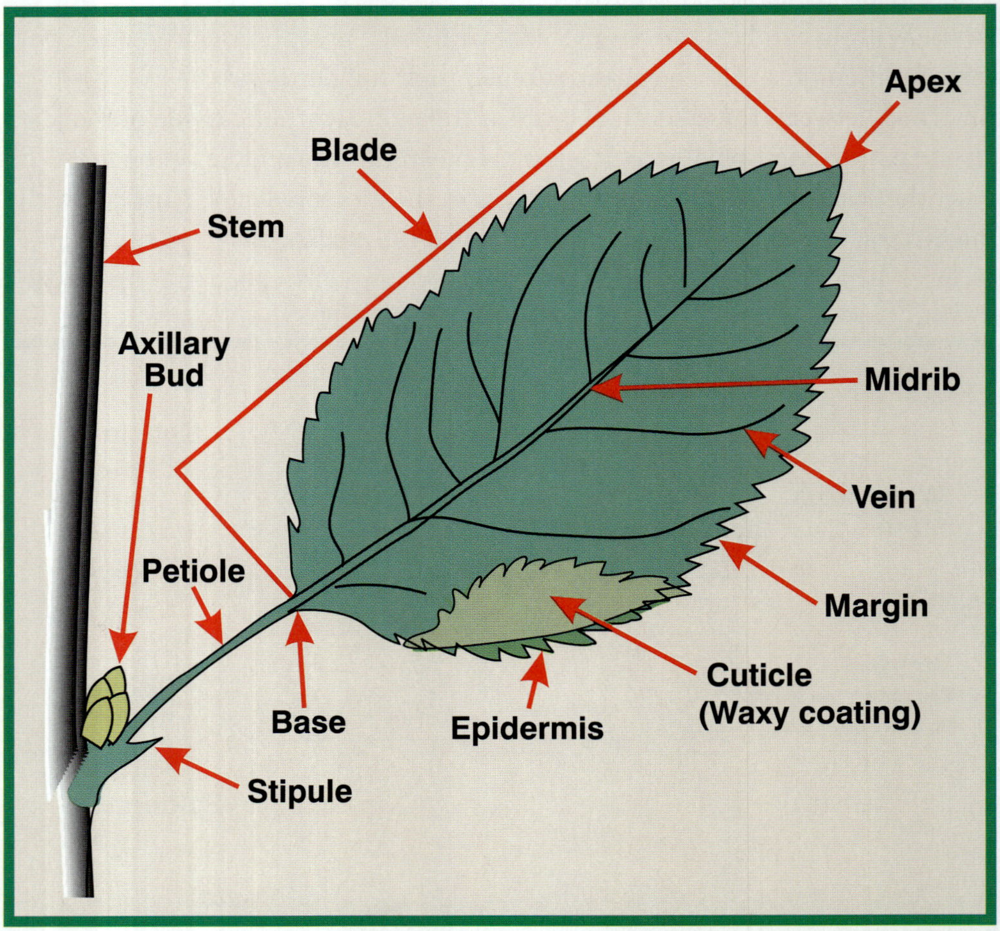

6–16. The simple leaf consists of a leaf blade and petiole.

6–17. Margins or the edges of leaves differ with the species.

6–18. Compound leaves have two or more leaflets and may be pinnately compound or palmately compound.

Leaves may be simple or compound. ***Simple leaves*** consist of a single leaf blade and a petiole. ***Compound leaves*** are made of a petiole and two or more leaf blades called leaflets. Compound leaves may be ***pinnately compound*** or ***palmately compound***.

Even when a plant is grown and used for its flowers, the foliage is important. A desirable plant will have leaves with an overall healthy appearance. They should have a good color and be free of disease and physical damage. Many floriculture crops are produced because of the ornamental value of the leaves. Most tropical houseplants are valued for their leaves.

6–19. Healthy leaves have good color and are free of disease.

6–20. Anthocyanins are responsible for blue flower colors. (Courtesy, Ball Horticultural Co.)

Leaf Color and Texture

Pigments determine the color of leaves. Pigments are coloring materials found in the cells. Most plants have a green color due to the pigment chlorophyll. One would expect this since **chlorophyll** absorbs light and is a key component of the photosynthetic process.

Other pigments in the plant produce a wide variation of yellows, blues, and reds. The pigment **carotene** produces orange-yellow colors. It plays a much smaller role in photosynthesis than chlorophyll. Carrots have a high carotene content. **Xanthophyll** is a pigment with colors that range from yellow to nearly no color at all. **Anthocyanins** are responsible for many of the blue, purple, and red colors. **Flavonols** produce yellow and cream colors. Both anthocyanins and flavonols have a greater influence on flower colors than leaf coloration.

The abundance of the green chlorophyll in leaves and stems usually masks the colors of other pigments. Other pigments are not seen unless the chlorophyll dies or is present in lower concentrations. A good example is the fall coloring of trees. Cold temperatures kill the chlorophyll allowing the other pigments to be seen. Plants with leaf colors other than green generally have a lower concentration of chlorophyll.

6–21. This Hosta exhibits variegated leaf patterns. (Courtesy, *Greenhouse Product News*)

Many floriculture plants are valued for the different color patterns of their leaves. These patterns are referred to as **variegation**. Examples of variegated plants include coleus, prayer plant, caladium, and Croton.

Modified leaves sometimes take on a petal-like appearance. They are referred to as **bracts**. The best example of a flowering plant with showy bracts is the poinsettia. The red, pink, or white structures are technically bracts, not petals. The small yellow structures in the center of the bracts are the actual flowers.

The texture of a leaf, which is so important to the aesthetic appeal, varies from species to species. Texture is influenced by many factors. The size of the leaf has a major impact on texture. Other factors are the margin of the leaf (entire leaves lack serrations or sinuses), the surface of the leaf (glossy verses rough), the venation of the leaf, and whether the leaf is simple or compound.

6–22. Poinsettias have showy bracts.

6–23. Leaf texture is influenced by a number of factors including the size of the leaf.

FLOWERS

The floriculture industry is based on the beauty of flowers. At times, it is easy to forget that flowers are plant organs with a specific function. ***Flowers*** are the reproductive organs of plants. A flower's function is to produce seeds through sexual reproduction. The seeds become the next generation of plants. Plant breeders work with flowers to produce new varieties of flowers for the industry.

6–24. Plant breeders develop new varieties of flowers. (Courtesy, Ball Horticultural Co.)

Parts of a Flower

The flower typically consists of four different parts. They are the sepals, the petals, the stamens, and the pistil. The flower stem is known as the ***pedicel***. The tip of the stem that holds the flower parts is the ***receptacle***.

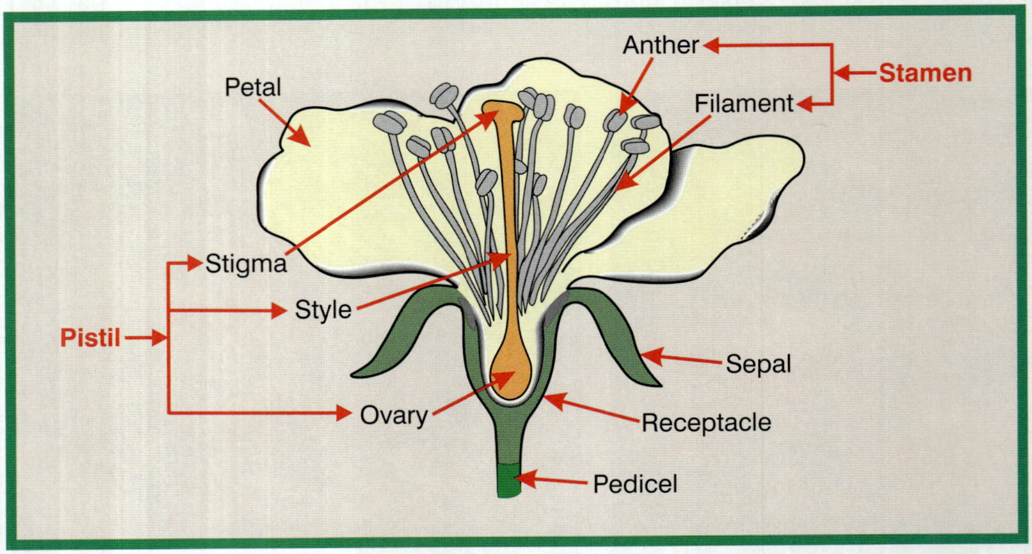

6–25. Parts of a flower.

The exterior of the flower consists of green, leaf-like structures called ***sepals***. The sepals fold back as the flower opens. Collectively, the sepals of a flower are called the ***calyx***. Hibiscus and the standard carnation are good examples of flowers with a calyx that surrounds the flower petals. The florist hydrangea is an unusual plant in that the sepals take on the colorful role that petals usually play.

Petals are located just inside the sepals. They are usually thought of first when discussing flowers. ***Petals*** appear leaf-like and are often very colorful. The function of brightly colored petals is to attract pollinators, such as insects and birds. Bees tend to be attracted to blue and violet flowers; hummingbirds tend to be attracted to red flowers. The collection of petals on a flower is referred to as the ***corolla***.

Together, the sepals and petals are called the ***perianth***. The perianth surrounds the reproductive parts of the flower. The functions of the perianth are to protect the reproductive parts of the flower and to attract pollinators. The perianth is not essential to the reproductive functions of the flower. With some plants, a clear distinction between petals and sepals cannot be made. In these cases, the parts of the perianth are called ***tepals***. Tulip flowers and many monocots have tepals.

The reproductive parts are found at the center of a flower. The ***stamens*** or male reproductive parts of a flower are arranged around the female parts of the flower. The stamen consists of a stalk called the ***filament*** and an anther. The ***anther*** produces and holds the pollen. The ***pollen*** grains contain the male sex cells. Lilies have pronounced stamens and produce an abundance of pollen.

6–26. Tulips have tepals.

6–27. The anthers on this lily flower are clearly visible.

The female part of the flower is the ***pistil***. The pistil has three main parts, the stigma, the style, and the ovary. The ***stigma*** is found at the end of the pistil and it has a sticky surface on which pollen can be caught. The neck of the pistil is referred to as the ***style***. The third part is the ***ovary***, which contains one or more ovules. The eggs are produced and seeds develop within the ***ovule***. As the seeds form, the ovary becomes a fruit. Ornamental pepper and Christmas cherry are potted plants grown for their attractive fruit.

Flowers that have sepals, petals, stamens, and a pistil are ***complete flowers***. A flower that lacks any one of these parts is called an ***incomplete flower***. Flowers that have both the male and female parts are said to be ***perfect flowers***. Those that lack one of the two sex structures are ***imperfect flowers***.

Numerical Plan of Flowers

Most flowers have a definite number of parts. Monocot flowers usually come in threes. They have three sepals, three petals, and usually three or a multiple of three stamens. This

6–28. The flower parts of gladioli and other monocots are in multiples of three. (Courtesy, Ball Horticultural Co.)

pattern can be seen in lilies, gladioli, iris, freesia, and orchids. Dicots are usually constructed in forms of four or five. These numbers are easiest to see in the sepals and petals.

Forms of Flowers

Flowers are borne on plant stems in one of two ways. Some plants have ***solitary flowers***, such as the tulip, narcissus, and rose. Other plants have flower clusters, known as an ***inflorescence***. An inflorescence is actually

6–29. A rose is a solitary flower.

6–30. An inflorescence consists of clusters of flowers on a stem. (Courtesy, California Cut Flowe Commission)

the branching system of the stem. The main stem of an inflorescence is known as the ***peduncle***. Pedicels that branch from the peduncle support individual flowers.

One way in which types of inflorescence are classified is by the sequence of flowering within the flower cluster. If the first flower to open is at the apex of the stem and the progression of flowering is downward or outward, the inflorescence is said to be ***determinate***. Some examples of determinate flowers include liatris, kalanchoe, and African violet. If the last flower to open is terminal on the main axis and the progression of flowering is inward or upward, the inflorescence is ***indeterminate***. Freesia, cineraria, snapdragon, and stock have indeterminate inflorescence.

6–31. When the first flower of an inflorescence to open is at the apex of the stem and the progression of flowering is downward, the inflorescence is said to be determinate. (Courtesy, California Cut Flower Commission)

6–32. This snapdragon is indeterminate because the terminal flower opens last. (Courtesy, *Greenhouse Product News*)

33. Baby's breath is a cyme. (Courtesy, California t Flower Commission)

Inflorescence Types

Cyme—A cyme takes on several forms, although they are usually a flat-topped inflorescence. A dichasium cyme bears a terminal flower on a peduncle with a pair of branches that produce lateral flowers. It may be simple or compound. A monochasium cyme has a terminal flower and, below it, one branch that produces a single lateral flower. The terminal

6–34. Liatris (spike) has flowers attached to the main stem without stalks.

6–35. Delphinium (raceme) has flowers attached to the main stem by short stalks. (Courtesy, Ball Horticultural Co.)

flower is the oldest. Alstroemeria is a cyme. Baby's breath is a compound dichasium cyme.

Spike—A spike is an elongated inflorescence with a central axis along which are sessile flowers. Gladioli and liatris are spike inflorescences.

Raceme—A raceme is an elongated inflorescence with a central axis along which are simple pedicels of more or less equal length. Examples of racemes include snapdragon, delphinium, Scotch broom, and stock.

Panicle—A panicle is an elongated inflorescence with a central axis along which there are branches that are themselves branched. Astilbe and begonia have a panicle inflorescence.

Corymb—A corymb is a short and broad, flat-topped indeterminate inflores-

6–36. Astilbe (panicle) has a highly branched inflorescence. (Courtesy, Ball Horticultural Co.)

cence having a main vertical axis and pedicels or branches of unequal length. The outer flowers open first. Yarrow is an example of a corymb.

Umbel—An umbel is an inflorescence having several branches arising from a common point. A simple umbel consists of flowers with single pedicels. Compound umbels have secondary branching in the form of pedicels at the end of a ray. Queen Anne's lace and amaryllis have umbels.

Spadix—A spadix is a spike with a thickened, fleshy axis, usually enveloped by a showy bract called a ***spathe***. Floriculture crops with a spadix include calla lily and anthurium.

6–37. Yarrow (corymb) has a short, flat-topped indeterminate inflorescence. (Courtesy, California Cut Flower Commission)

6–38. Amaryllis (umbel) has flower stalks that arise from a common point at the tip of the stem.

Catkin—A catkin is a spike, raceme, or cyme composed of unisexual flowers without petals and falling as a unit. Catkins are found on willows, alders, oaks, and birch.

6–39. Anthurium (spadix) has a spike with a thickened, fleshy axis, enveloped by a showy bract.

Head—A head is a rounded or flat-topped cluster of sessile flowers. Head inflorescences of the aster family resemble single flowers. These consist of centrally grouped flowers called disc flowers encircled by ray flowers. Plants with head inflorescence are very important to the floriculture industry. Some common plants that have a head inflorescence are gerbera daisy, chrysanthemum, sunflower, marigold, dahlia, strawflowers, and cineraria.

-40. Willows (catkins) have an upright inflorescence. (Courtesy, Ken Butler)

6–41. Sunflower (head) has an inflorescence made up of a cluster of sessile flowers.

REVIEWING

MAIN IDEAS

Plants have four main structures including the roots, stem, leaves, and flowers. Each structure performs specific functions for the plant.

Roots support the above ground growth of a plant. They absorb water and minerals, and they anchor the plant. Root hairs increase the surface area for water absorption. Adventitious roots grow from the stem. Roots also store food for the plant. Growers understand the importance of maintaining a healthy root system.

Stems hold the leaves and flowers upright. Materials within the plant are transported up through the xylem and down through the phloem. Buds along the stem protect undeveloped leaves, stems, and flowers. The terminal bud at the apical meristem produces auxins that inhibit the growth of lateral buds. Underground stem structures that store food and water include bulbs, corms, rhizomes, and stem tubers.

Leaves are organs with the responsibility of making food. Most photosynthetic activity occurs in the mesophyll layer of cells. The stomata are openings in the leaf controlled by guard cells. They allow the exchange of gases. Leaves may be simple or compound. Pigments in the cells give the leaves color. Chlorophyll, a green pigment, is in abundance and converts light into stored energy.

A flower is the reproductive organ of a plant. The flower typically consists of four different parts. They are the sepals, the petals, the stamens, and the pistil. A single flower on a flower stem or peduncle is called a solitary flower. An inflorescence is a flower cluster. The structure of an inflorescence allows it to be classified.

QUESTIONS

Answer the following questions using complete sentences and correct spelling.

1. What are the major parts of a plant?
2. What are the functions of a root?
3. What are the functions of a stem?
4. What does the xylem do for a plant?
5. What is the function of the leaf?
6. What role does pigment play in a plant?
7. What is the function of the flower?
8. What parts make up a complete flower?
9. What is the difference between a determinate and indeterminate flower inflorescence?
10. How do inflorescence types differ?

EVALUATING

Match the term with the correct definition. Write the letter of the term on the line provided.

a. phloem
b. root hairs
c. stoma
d. stamen
e. corolla
f. determinate
g. stomata
h. lateral bud
i. adventitious root
j. calyx
k. bract
l. petiole

______1. All of the sepals of a flower.

______2. An inflorescence whose first flower to open is at the apex of the stem and the progression of flowering is downward or outward.

______3. All of the flower petals.

______4. Male part of a flower.

______5. Thin roots that absorb water and minerals.

______6. The stem of a leaf.

______7. Pore or opening in the leaf.

______8. Conductive tissue through which food moves.

______9. Contain undeveloped leaves, flowers, or stem and are located along the stem.

______10. Sticky surface at the end of the pistil.

______11. Modified leaves that take on a petal-like appearance.

______12. Roots that begin growth from the stem or a leaf.

EXPLORING

1. Collect leaves from different types of plants. Place them under a microscope. Examine the stomata. Tear the leaves and look at the interior cells. Are there differences between the plants?

2. Sow some radish seeds on a moist paper towel. Put the towel in a plastic bag. Remove the plastic bag in three days and look at the roots. Identify the primary root and the root hairs. Examine the root hairs under magnification.

3. Look at a flower in the garden. See if you can identify the major parts of the flower. Remove the sepals and petals to get a better view of the pistil.

CHAPTER 7

Life Processes and Growth Regulators

OBJECTIVES

This chapter is about the life processes and growth regulators of plants. It has the following objectives:

1. Explain photosynthesis in simple terms.
2. Describe cellular respiration and its importance in plant growth.
3. Identify plant growth regulators (PGR).
4. Recognize the functions of plant hormones.
5. Relate the application of synthetic growth regulators to the floriculture industry.

TERMS

abscisic acid
cellular respiration
cellulose
chloroplasts
cytokinin
enzymes
ethylene
etiolation
gibberellins
gravitropism
growth retardants
hormones
indoleacetic acid (IAA)
indolebutyric acid (IBA)
metabolism
naphthaleneacetic acid (NAA)
phototropism
plant growth regulators (PGR)
starch
synthetic growth regulators

7–1. Beautiful plants are a result of complex life processes.

DOES plant life fascinate you? Do you find it interesting that small seeds can produce beautiful bedding plants? Do you enjoy the great variety of floral products available to you, in all their colors? It has been said that people with an interest in gardening and horticulture love life. They love the life processes involved in plant growth. However, the greatest appreciation for plants is usually associated with their aesthetic qualities.

The challenge for those in the floriculture industry is to deliver floral products people find attractive. After determining what is considered attractive, growers must be able to produce the floral products when people want them. To do so, growers must first understand the life processes of plants. Then, growers need to manipulate growing conditions for optimum growth and development. They may also need to use materials that regulate growth to produce high quality crops.

LIFE PROCESSES OF PLANTS

7–2. Photosynthesis takes place in green plants.

Plants are living organisms. As such, they have complex chemical processes that direct growth and development. Photosynthesis and respiration are two major processes. These and other life processes are regulated by hormones and conducted by enzymes.

PHOTOSYNTHESIS

Green plants have the unique ability to produce their own food. The process is known as photosynthesis. During photosynthesis, simple molecules are converted into more complex molecules. The complex molecules store energy in the form of chemical energy for later use. The photosynthetic process can be broken down into two stages, the light reaction and CO_2 fixation.

The process of producing food begins with the light reaction phase. Light energy is trapped and used to fuel the photosynthetic process. The harnessing of light energy is made possible by pigments found in chloroplasts. ***Chloroplasts*** are specialized organelles within the individual plant cells. As light enters the chloroplast, chlorophyll, carotene, and xanthophyll pigments absorb the energy.

Chlorophyll, the primary pigment involved in the manufacture of food, uses the light energy to make the high-energy compounds ATP and NADPH that power reactions in the cells. Chlorophyll uses the energy to split water molecules, H_2O, during the first phase of the photosynthetic reaction. Oxygen atoms from the water molecules bond to form O_2. The O_2 escapes through the stomata into the atmosphere. Hydrogen atoms from the water molecule are incorporated into sugar molecules as the process continues in CO_2 fixation.

During the CO_2 fixation phase of the photosynthetic reaction, energy in the form of ATP and NADPH is used to form sugars. Carbon dioxide enters through the stomata and is available to the chlorophyll in the cells. The CO_2 combines with hydrogen from the water molecule. Glucose and fructose ($C_6H_{12}O_6$) are assembled. Although they both have the same formula, their molecular structures differ.

7–3. Chloroplasts contain chlorophyll and other pigments that trap light energy.

Glucose and fructose fuel plant growth. They are the source of energy for the plant's life processes. All live cells in the plant benefit as the sugars are transported through the phloem to the rest of the plant. The simple sugars can also be combined to form a more complex sugar, sucrose ($C_{12}H_{22}O_{11}$).

The sugar molecules may be processed further to form starch and cellulose. These are huge molecules resulting from the bonding of thousands of glucose molecules. ***Starch*** serves as the principle way in which food is stored for plants. When needed, it is easily broken down into glucose or converted into other plant products by plant enzymes. The ***cellulose*** is applied to

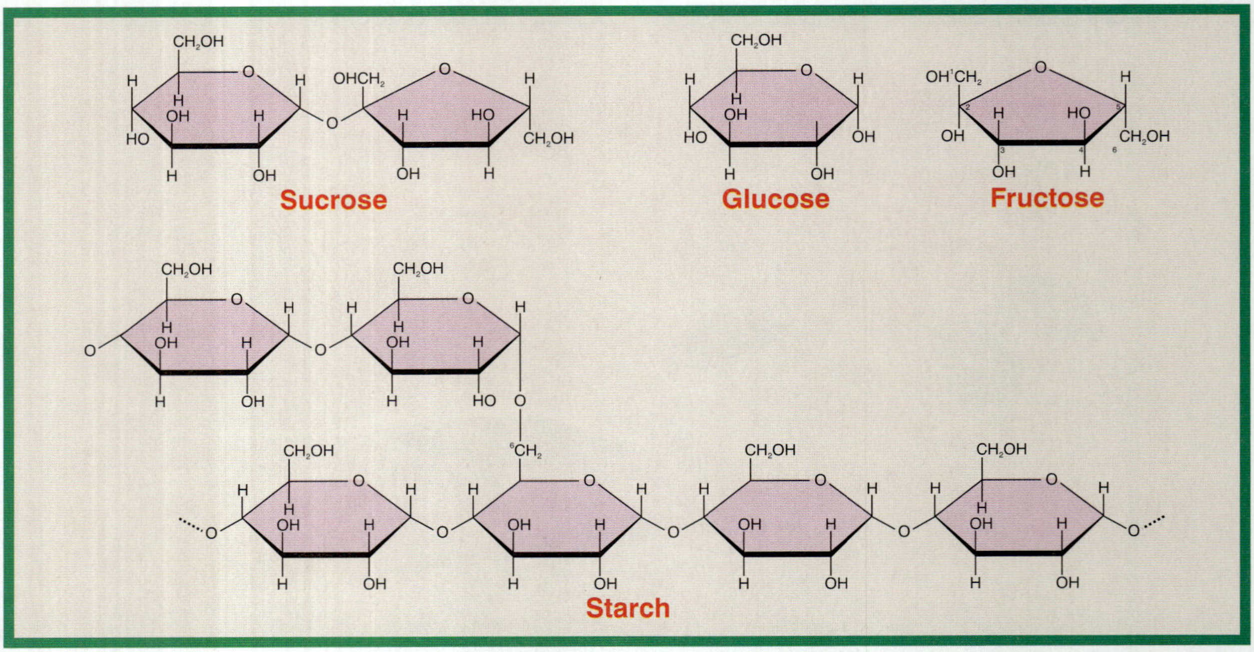

7–4. Sugars are produced in photosynthesis.

cell walls for strength and rigidity. Once sugars are converted to cellulose, they are not reclaimed for other purposes.

A simple chemical equation for photosynthesis follows:

$$6\,CO_2 + 12\,H_2O \xrightarrow[\text{chlorophyll}]{\text{light}} C_6H_{12}O_6 + 6\,O_2 + 6\,H_2O$$

7–5. Supplemental lighting is being used with these chrysanthemum cuttings in December.

It is to the advantage of greenhouse growers to maximize the photosynthetic activity in their crops. High rates of photosynthesis contribute to healthier plants. Also, crops develop more rapidly when photosynthetic rates are high. Therefore, growers strive to maintain optimum levels of limiting factors, such as water and light. They carefully monitor watering. Supplemental lighting may be used. Also, some greenhouse operations are equipped with CO_2 burners to raise CO_2 levels and, thus, speed plant growth.

Table 7-1. A Comparison of Photosynthesis and Respiration

Photosynthesis	Respiration
CO_2 and H_2O are used. Food and O_2 are produced. Energy from light is trapped by chlorophyll and food. ATP is produced through photophosphorylation. Only cells containing chlorophyll carry out photosynthesis. Photosynthesis occurs only in light.	O_2 and food are used. CO_2 and H_2O are produced. Every living cell carries out respiration. ATP is produced through oxidative phosphorylation. Respiration occurs in both light and dark.

CELLULAR RESPIRATION

The chemical process known as ***cellular respiration*** is the reverse of photosynthesis. In cellular respiration, sugars made in photosynthesis are broken down into simpler molecules. In the process of breaking the chemical bonds, energy is released. The energy is applied toward growth and development of the plant.

7–6. Growth and development are results of cellular respiration.

Cellular respiration is extremely important in the growth and development of plants. All living plant cells respire or use energy to live and function, including humans. To germinate, a seed needs to break down the food that has been stored in the cotyledons or endosperm. For root cells to function, they need to convert sugars to energy.

Cellular respiration involves sugars produced in the photosynthetic process along with oxygen and water. In fact, oxygen is a critical ingredient to the reaction. This explains why plant roots need a medium that is well aerated. In the reaction, chemical energy is released when the molecular bonds of the sugar molecules are broken. The extracted energy, ATP, drives a variety of chemical reactions in the cell. By-products of the reaction include carbon dioxide and water.

A simple equation for cellular respiration follows:

$$C_6H_{12}O_6 + 6\,O_2 + 6\,H_2O \longrightarrow 12\,H_2O + 6\,CO_2 + \text{Energy}$$

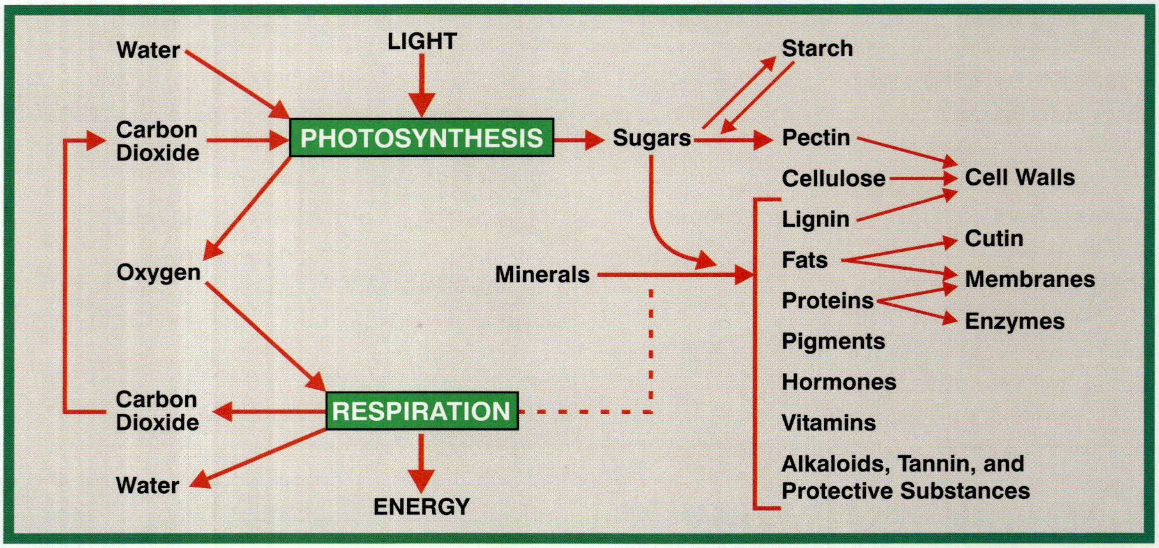

7–7. Plant metabolism flow chart.

Actual plant growth, fueled by cellular respiration, takes place primarily at night when photosynthesis is shut down. With signals from hormones, ***enzymes*** or chemical activators are produced. Each enzyme has a specific job. With split-second timing, they break down sugars and recombine them with nitrogen and other minerals. Many complex molecules are produced. They include starches, pectin (to bind cells), lignin (a tough, durable substance), cellulose, lipids (fats), proteins, pigments, hormones, vitamins, and alkaloids and tannins (materials that protect plants from pests and diseases).

7–8. Soil temperatures can be monitored with a soil thermometer.

All of the chemical reactions in a plant fall under a term, ***metabolism***. The speed at which chemical reactions occur is influenced by temperature. Plant metabolism is slowed in cool or cold temperatures and is more rapid in warmer temperatures. Growers adjust greenhouse temperatures to control the quality and timing of floriculture crops. Florists keep cut flowers in coolers to slow metabolism. Some greenhouse crops can be placed in coolers to slow their development.

7–9. These florist azaleas are being moved to a cooler to slow their metabolism.

GROWTH REGULATORS

Naturally occurring or synthetic chemicals regulate all growth and development of plants. These chemicals are known as ***plant growth regulators (PGR)***. They may promote growth, inhibit growth, or modify plant growth and development. Natural chemicals produced by plants to regulate growth are called ***hormones***. Other chemical compounds that regulate growth, but are not produced by plants, have been discovered by plant physiologists. These compounds are called ***synthetic growth regulators***.

PLANT HORMONES

There are five different hormones produced in plants: auxins, gibberellins, cytokinins, ethylene, and abscisic acid. Each hormone promotes many different plant responses, and each is effective in very low concentrations. Each interacts with the other hormones in complex ways to produce plant responses. Ultimately, several different hormones may affect a particular plant response.

Auxins

Auxins are a group of compounds that stimulate certain plant responses. The major plant response influenced by auxins is cell elongation. Auxins are produced in the apical meristem of a plant's stem and migrate down the stem. They move from one cell to the next rather than through the xylem or phloem. The primary auxin is ***indoleacetic acid (IAA)***.

7–10. Auxins are responsible for phototropism, the ability of a plant to bend toward light.

Auxins are responsible for ***phototropism***, or the ability of a plant to bend toward a light source. The way this works is the auxins move down the shaded side of a stem causing those cells to elongate more than the cells on the bright side of the stem. The effect is a stem bending toward the light.

7–11. A plant's response to gravity that results in a stem growing upward is known as gravitropism.

Auxins cause a plant's response to the force of gravity. When a stem is placed on its side, auxins move to the lower side of the stem. Cells on the lower side elongate and the stem curves upward. This response to gravity is known as ***gravitropism*** or geotropism. It is easy to see in plants that have tipped over in a greenhouse.

Auxins play a role in apical dominance. As they move down the stem from the apical meristem, they inhibit growth of side shoots. Lateral buds farthest away from the apical meristem are influenced the least by the auxins. As a result, they develop sooner than buds closest to the tip. Pinching off the apical meristem stops the flow of auxins down the stem, allowing side shoots to develop. Pinching is a common practice used to produce bushy, well-branched crops. A few greenhouse crops that are pinched include poinsettias, zonal geraniums, and chrysanthemums.

Root development on the cuttings of some plants is promoted naturally by auxins. Many foliage plants produce plenty of auxin to make rooting easy. Other plants need to be treated with a synthetic auxin to produce roots.

Gibberellins

Gibberellins are produced in stem and root apical meristems, in seed embryos, and in young leaves. They induce stem cell elongation and cell division. Gibberellins play a key role in stimulating the development of flowers. They also have an important role in the production of enzymes during seed germination.

Gibberellins cause internodes to stretch. This action is related to the intensity of light. Under high light intensity, the hormone's effect is inhibited. However, if light levels are low, gibberellins cause the internodes to lengthen. By this action, leaves are raised to a position whereby they can be in position to capture light. A common problem in the greenhouse occurs when plants are spaced too closely to one another. The plants shade one another. Stretching takes place that results in less compact plants, weaker stems, and ultimately a crop of lesser value. Plants grown in low light or darkness develop a severe case of stretching known as ***etiolation***.

7–12. Gibberellins cause internodes to stretch.

Gibberellic acid is sometimes used as a growth regulator. One use is as a substitute for cold treatment for azaleas and hydrangeas. The treated azaleas and hydrangeas produce flowers without actually having gone through a cold period. Gibberellic acid is also used to increase the peduncle (flower stem) length on the pompon chrysanthemum and the stem elongation on the standard chrysanthemum, fuchsia, and geranium.

7–13. Foliar application of gibberellic acid is sometimes used to induce flowering of azaleas and hydrangeas in place of a cold treatment.

7–14. When the source of auxins is removed by pinching, cytokinins promote the growth of lateral shoots.

Cytokinins

Cytokinins are responsible for cell division and differentiation. They are produced in the roots and transported throughout the plant through the xylem. Cytokinins play a role in apical dominance of intact plants. While auxins inhibit the growth of lateral shoots, cytokinins promote lateral shoots. In roots, the situation is reversed. Auxins promote the branching of roots, and cytokinins inhibit branching.

Cytokinins are important in plant propagation. In tissue culture or micropropagation, cytokinins must be added to the medium for cell division to occur. Also in tissue culture, cytokinins work with auxins causing cells to form plant organs. Medium with a high level of cytokinins in relation to auxins induces shoot formation. A low level of cytokinins in relation to auxins promotes root formation.

Cytokinins, and to a lesser extent auxins and gibberellins, delay aging or senescence of plants. Cells of plants age as they do in other organisms. Once the aging process in a plant begins, it is irreversible. Senescence is accelerated in portions of a plant removed from the main plant. Cut flowers are a good example. Their source of cytokinins, produced in the roots, is lost, and their lives are shortened. Cut flowers are sometimes sprayed with cytokinins to delay senescence and to extend their vase life.

Ethylene

Ethylene is a colorless gas produced in the nodes of stems, ripening fruits, and dying leaves. It causes the thickening of stems. It also speeds the aging of plant parts, particularly fruit. As cells age, chlorophyll breaks down, cell membranes weaken, and cell walls soften. Ripening fruit produces ethylene, which triggers the production of more ethylene, speeding the ripening process.

7–15. Cytokinins are added to tissue culture medium to promote cell division.

Ethylene gas is used and avoided depending on the floriculture operation. Growers for years placed ripened fruit with florist hydrangeas as part of the production schedule. The ethylene caused hydrangeas to drop their leaves before the plants were placed in a cooler. If cut flowers could cringe, they would in the presence of ethylene. Ethylene shortens the life of the flowers. Therefore, they should never be stored with ripe fruit or decaying leaves that might give off ethylene.

Abscisic Acid

Abscisic acid is a growth-inhibiting hormone. It is largely responsible for seed dormancy. The influence of abscisic acid over seed dormancy decreases during winter,

7–16. Abscisic acid causes stomata to close during times of drought.

while levels of gibberellins and cytokinins increase. The re-proportioning of the hormones leads to germination in the spring. Abscisic acid also contributes to senescence along with ethylene.

Table 7–2. The Affects of Plant Hormones on Growth and Development

Physiological Activity	Auxin	Gibberellin	Cytokinin	Ethylene	Abscisic Acid
Seed Germination			Promotes		Inhibits
Growth of Seedling into Mature Plant	Cell elongation	Cell elongation and division	Cell division and differentiation		
Apical Dominance	Inhibits lateral bud development		Promotes lateral bud development		
Flower Initiation			Stimulates flowering in some plants		
Fruit Development and Ripening	Development	Development		Promotes ripening	
Leaf Abscision	Inhibits		Inhibits	Promotes	
Winter Dormancy		Breaks dormancy			Promotes dormancy
Seed Dormancy		Breaks dormancy			Promotes dormancy

Another role abscisic acid plays becomes most apparent when a plant is under stress. It is abscisic acid that causes stomata to close during times of drought. By doing so, the amount of water lost from plant tissues is reduced.

SYNTHETIC GROWTH REGULATORS

A number of synthetic compounds are utilized in the floriculture industry. They range from materials that promote growth to those that inhibit growth. With some plants, some synthetic growth regulators are more effective than others.

7–17. These zonal geranium cuttings will be treated with rooting hormones before sticking. (Courtesy, Ball Horticultural Co.)

Rooting Compounds

Synthetic auxins are also used in the horticulture industry to promote rooting of cuttings. Cuttings are treated with synthetic growth regulators to increase the number of cuttings that form roots, to speed rooting, to increase the number and quality of roots, and to increase the uniformity of the roots.

The two synthetic root-promoting materials most widely used are ***naphthaleneacetic acid (NAA)*** and ***indolebutyric acid (IBA)***. Of these, IBA is the best material for general use. It can be used with a wide variety of plants and it is nontoxic over a wide range of concentrations. These materials are available in a liquid formulation or mixed with talc.

Growth Retardants

Growth retardants are widely used in the greenhouse industry. These chemicals inhibit the action of gibberellins on stem elongation. As a result, plants are more compact. Compact plants are more attractive, often greener, and easier to transport. Although

7–18. Growth retardants are used to produce compact plants such as this lisianthus. (Courtesy, Ball Horticultural Co.)

growth is slowed, flowers tend not to be affected. The major growth retardants and some of their uses follow:

- Ancymidol (A-Rest) is used to control the height of bedding plants, zonal geraniums, foliage plants, eustoma, and monocot plants, including lilies.
- Paclobutrazol (Bonzi) is used to control the height of tulips, gloxinia, gerbera daisies, and freesias.
- Chlormequat (Cycocel) is an effective retardant on poinsettias, zonal geraniums, seed geraniums, and hibiscus.
- Ethephon (Florel or Pistil) is used with geraniums grown from cuttings to encourage branching (mostly stock plants), and with hyacinths, daffodils, and foliage plants to control height.
- SADH (Alar or B-Nine) is the most important growth retardant. It is used to control the height of many crops including bedding plants, zonal geraniums, perennials, chrysanthemums, gloxinia, hydrangeas, gerbera daisies, exacum, eustoma, and kalanchoe.
- Uniconizole (Sumagic) is used to control the height of perennials and lilies.

As with any chemical, certain precautions need to be followed when working with growth regulators. Suggestions for applying growth regulators follow:

- Read the label and follow instructions as to the use of the material.
- Wear appropriate safety gear when mixing and during application.
- Apply the proper growth regulator to the proper plant at the proper time and at the proper application rate.

7–19. Growth regulators produce different results when used in different concentrations. (Courtesy, *Greenhouse Product News*)

- Be sure the plant is well established with a good root system before applying any growth regulator.
- Apply the material uniformly either as a foliar spray or by thoroughly soaking the soil mass when drenching.
- Although more difficult to apply, a single drench of Cycocel or A-Rest will generally be more effective than spray treatments.
- If possible, split a full-strength application into two half-strength applications.

REVIEWING

MAIN IDEAS

One of the two major life processes in plants is photosynthesis. This chemical reaction has two phases. The first phase involves the trapping of light energy by pigments in the chloroplasts. One of those pigments, chlorophyll, converts the light energy to high energy ATP and NADPH and breaks water molecules. In the second phase, sugars are manufactured. Glucose and fructose are stored energy for the plant. They can be processed further to be sucrose, starch, and cellulose.

The other important chemical process in plants is cellular respiration. Cellular respiration is the opposite of photosynthesis in that energy is released as the stored sugars are broken down. Cellular respiration takes place primarily at night when photosynthesis is shut down. Enzymes, produced with signals from hormones, make complex molecules that include starches, pectin, lignin, cellulose, lipids, proteins, pigments, hormones, vitamins, and alkaloids and tannins. All of the chemical processes are a plant's metabolism.

Plant growth regulators (PGR) regulate all growth and development of plants. They may promote growth, inhibit growth, or modify plant growth and development. Natural chemicals produced by plants to regulate growth are called hormones. Plants produce five hormones, auxins, gibberellins, cytokinins, ethylene, and abscisic acid. Each hormone promotes many different plant responses, and interacts with the other hormones in complex ways to produce plant responses.

Other chemical compounds that regulate growth but are not produced by plants are called synthetic growth regulators. Synthetic root-promoting materials are used to increase the number of cuttings that form roots, to speed rooting, to increase the number and quality of roots, and to increase the uniformity of the roots. Growth retardant chemicals result in compact plants.

QUESTIONS

Answer the following questions using complete sentences and correct spelling.

1. How is chlorophyll involved in the photosynthetic reaction?
2. How are sugars formed in CO_2 fixation?
3. How does cellular respiration differ from photosynthesis?
4. What are products resulting from enzyme activity?
5. How do hormones and synthetic growth regulators differ?
6. What are the major functions of auxins?
7. What are the major functions of gibberellins?
8. What are the major functions of cytokinins?
9. What are the major functions of ethylene?
10. What are the major functions of abscisic acid?
11. What are some uses of synthetic growth regulators?

EVALUATING

Match the term with the correct definition. Write the letter of the term on the line provided.

a. chloroplasts
b. gibberellins
c. hormones
d. starch
e. growth retardants
f. enzymes
g. plant growth regulators (PGR)
h. senescence
i. metabolism
j. phototropism

______1. Serves as the principle way in which food is stored for plants.

______2. Chemical activators.

______3. Produced in stem and root apical meristems, they induce stem cell elongation and cell division.

______4. Aging of plants.

______5. All of the chemical reactions in a plant.

______6. Natural chemicals produced by plants to regulate growth.

______7. The ability of a plant to bend toward a light source.

______8. Widely used in the greenhouse industry to inhibit the action of gibberellins on stem elongation.

______9. Naturally occurring or synthetic chemicals that regulate all growth and development of plants.

_____10. Specialized organelles within the individual plant cells.

EXPLORING

1. Run an experiment using synthetic rooting hormones. Take stem cuttings. Treat some of the cuttings with different concentrations of rooting hormone. Leave other cuttings untreated. Place the cuttings under conditions that induce rooting. Monitor the rooting and determine what effect the treatments might have had on the outcome.

2. Visit a local greenhouse and investigate their practices. What do they do to maintain high levels of photosynthetic activity? Ask what advantage it is for them to produce a crop quickly. Do they use growth regulators? If so, how do they use them and on which crops?

CHAPTER 8

Plant Propagation

OBJECTIVES

This chapter is about plant propagation. It has the following objectives:

1. Describe the importance of hybridizing plants for the greenhouse industry.
2. Describe the importance of propagating plants for the industry.
3. Explain methods of sexual propagation.
4. Explain methods of asexual propagation.
5. Describe the role of tissue culture and biotechnology in the greenhouse industry.

TERMS

air layering
asexual propagation
budding
bulbils
bulblets
callous
chromosomes
clone
cormel
cotyledon
cross-pollination
cutting
deoxyribonucleic acid (DNA)
diploid
division
double fertilization
embryo
endosperm
explant
fertilization
genes
genetic engineering
germination
grafting
haploid
hybrid
hybridization
layering
leaf-bud cutting
leaf cutting
leaf-petiole cutting
plantlet
pollination
polyploid
rootstock
scarification
scion
scooping
scoring
seed
seed coat
self-pollination
separation
sexual propagation
stem cutting
sticking cutting
stock plant
stratification
tetraploid
tissue culture
triploid
viability
vigor
zygote

8–1. Millions of plants are propagated annually in the floriculture industry. (Courtesy, University of Illinois Extension)

TAKE a cutting of a *Philodendron* and put it in water. Throw some marigold seeds in the garden and watch them grow. Starting new plants is fun. In fact, few aspects of the floriculture industry are as enjoyable as plant propagation. Producing new plants through propagation is constructive. It gives one a feeling of achievement. In a sense, propagating plants is creative. Maybe you have witnessed an older family member start plants in the home. People through the years have gained satisfaction from propagating plants in their own homes.

More importantly, plant propagation is a major component of greenhouse production. All cut flowers, potted flowering plants, and foliage plants begin as seeds or are propagated by some part of the plant. In effect, millions of plants are propagated annually throughout the industry. Some people even specialize in the propagation of plants.

PLANT PROPAGATION

8–2. Sexual reproduction involves seeds. (Courtesy, Ball Horticultural Co.)

Plant propagation is simply increasing the number of plants. It is accomplished in one of two general ways. The plants can be produced from seed, which is referred to as sexual reproduction, or they can be produced from parts of a plant by means of asexual reproduction. The methods of propagation and techniques used vary with the plant and the preference of the grower.

SEXUAL PROPAGATION

Sexual propagation involves flowers and seeds. In sexual propagation, flowers are pollinated and eggs fertilized. It involves the fusion of sex cells, sperm, and egg. Since both cells contribute genetic material, the resulting offspring is genetically different from the parent plants.

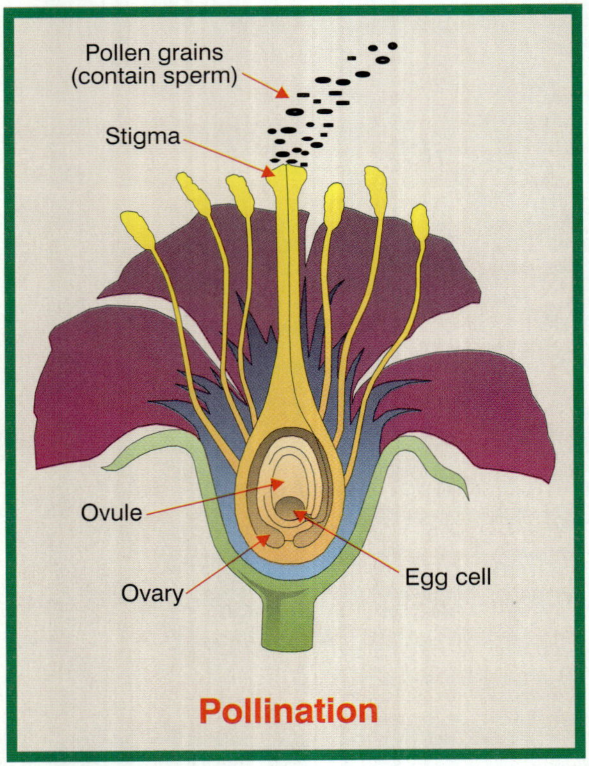

8–3. The transfer of pollen to the stigma is called pollination.

Pollination

The transfer of pollen to the female part of a flower is called **pollination**. In nature, flowering plants accomplish pollination a number of ways. Colorful, scented flowers attract birds, insects, bats, and other animals. These creatures unknowingly pick up pollen from the anthers and, when they visit another flower, deposit the pollen on the stigma. The plant rewards pollinators with sugary nectar. Other plants rely on the wind to transfer pollen to the stigma. The force of the wind physically moves pollen from one flower to another.

Since there is no need to attract pollinators, these plants do not produce colorful flowers with large petals, scents, or nectar.

When the pollen of a plant pollinates a flower on the same plant, it is called ***self-pollination***. Some plants have this ability; others do not. When the pollen of a plant pollinates the flower of another plant, it is said to be ***cross-pollination***.

8–4. Cross-pollination involves flowers of two different plants.

Most of the flower and plant varieties used in the floriculture industry today are the result of cross-pollination by plant breeders. The plant breeders select plants for outstanding characteristics, such as flower color and disease resistance. Then they collect pollen from one plant and transfer it to the flower of another. Their hope is to have the outstanding characteristics expressed in the offspring. This process is called ***hybridization***. The offspring of two plants of the same or related species that differ genetically is said to be a ***hybrid***. The advantage of hybridizing is desired traits of different parent plants can be combined in the offspring.

8–5. Most floriculture crops are hybrids.

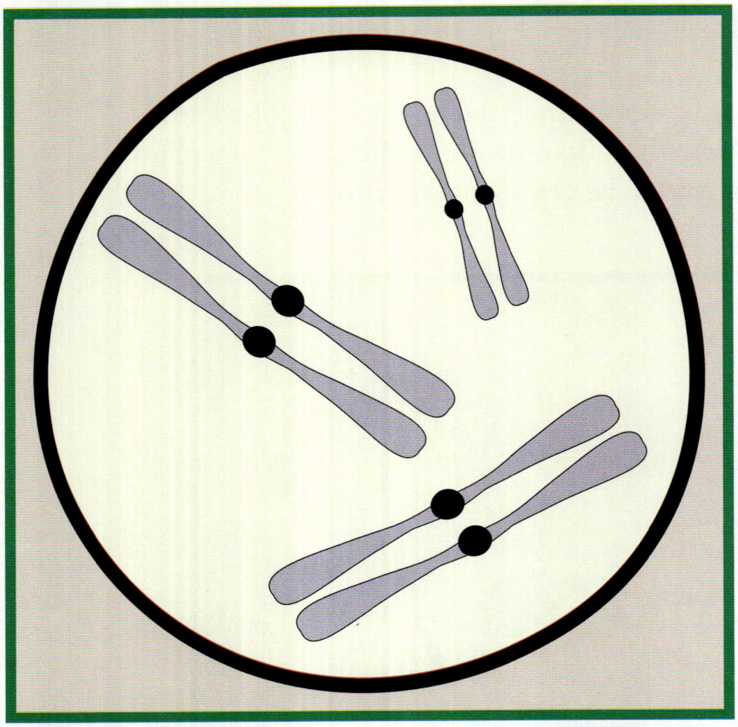

8–6. A normal cell with a double set of chromosomes is diploid.

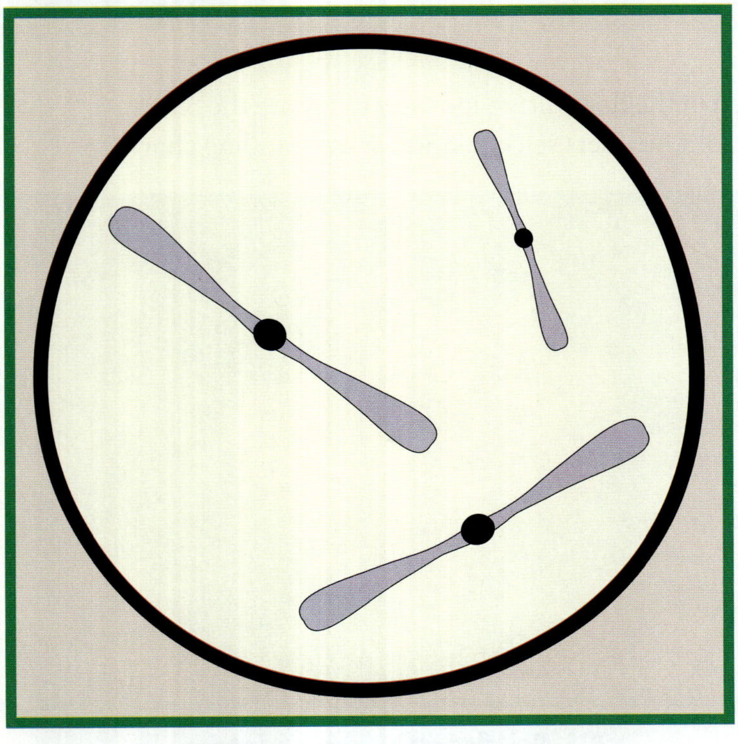

8–7. Reproductive cells are haploid.

Plant Genetics

An understanding of genetics is necessary when hybridizing plants. Genetic information is stored in every cell of a plant in long molecular chains made of ***deoxyribonucleic acid (DNA)***. Segments of the DNA, called ***genes***, code for life processes and the appearance of a plant. For instance, genes in a petunia plant may code for a red flower. Genes in that petunia plant also tell it when to flower and how many petals to make. An individual plant may have 100,000 genes informing it what to do and how to look.

Genes are arranged in a set of ***chromosomes***. The nuclei of normal cells contain a double set of chromosomes and are said to be ***diploid*** (2n). Reproductive cells, sperm and egg, have a single set of chromosomes and are said to be ***haploid*** (1n). When fertilization occurs, both the sperm and the egg contribute a single set of chromosomes. The resulting zygote ends up with a normal double set. In this way, traits from each of the parent plants may be passed on to the offspring.

In the floriculture industry, many plants have more than the normal double set of chromosomes. They are said to be ***polyploid***. Plants that have three sets of chromosomes are ***triploid*** (3n). Those with four sets of chromosomes are ***tetraploid*** (4n). Polyploids, endowed with extra sets of chromosomes, produce vigorous growth. They also have larger

8–8. The two larger flowers on the left in this photograph are tetraploid Stella de Oro daylilies. The ones on the right are diploid Stella de Oro daylilies. (Courtesy, Klehm Nursery)

flowers, leaves, and fruit than their diploid counterparts. Consequently, they are popular choices for the garden. Many daylily, dahlia, and gladiola varieties are polyploid. Polyploid plants frequently are sterile and cannot be propagated sexually. Therefore, they must be propagated through asexual means.

Fertilization

Once the pollen lands on the stigma, it grows a thin pollen tube down the style to the ovary. The cell within the grain of pollen divides to form two sperm nuclei. These travel down the pollen tube to an ovule that holds the egg. The ovule of the flower eventually becomes the seed. ***Fertilization*** is accomplished when a sperm fuses with an egg.

Fertilization in flowering plants is unlike fertilization in any other living organism. This is because both sperm nuclei in the pollen grain are involved in fertilization. This process, unique to flowering plants, is actually a ***double fertilization***.

The first fertilization occurs when a sperm fuses with the egg. The fertilized

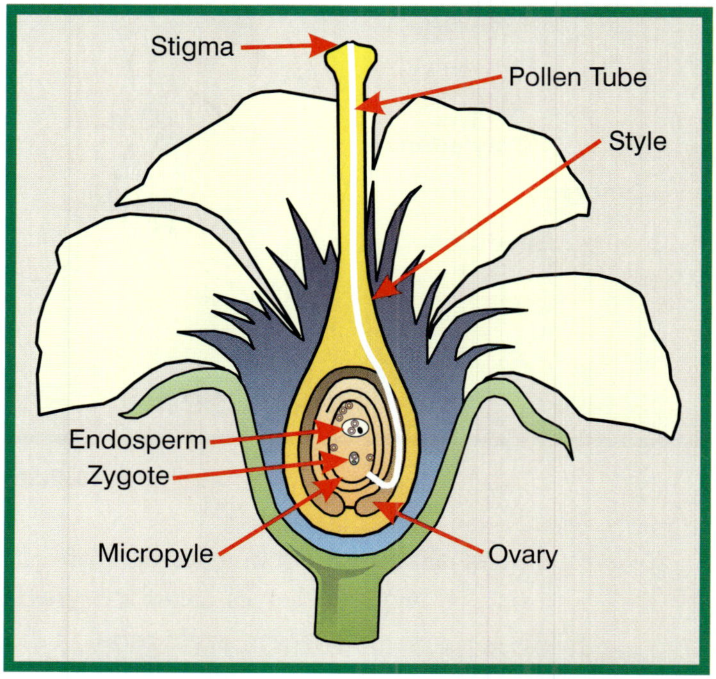

8–9. Fertilization occurs when the sperm fuses with the egg to form a zygote.

egg is called a ***zygote***. In this process, the sperm carries genetic material from the male part of the flower. The egg contains genetic material from the female part of the flower. The resulting zygote contains genetic material from both the male and female parts of a flower. Through cellular division, the zygote becomes the ***embryo*** or immature plant.

In the second fertilization, a sperm nucleus fuses with two nuclei in the ovule. It develops into the endosperm. The ***endosperm*** is tissue in which food is stored for the embryo.

Seeds

Seeds are truly a wonder of nature. They contain everything necessary for the growth and development of a new plant under the right conditions. ***Seeds*** have an embryo and a source of stored food contained within a seed coat. The ***seed coat*** is a protective shell that protects the embryo and the endosperm from drying and from physical injury. The seed coat also plays an important role in determining when outside conditions are right for ***germination***, the beginning of growth.

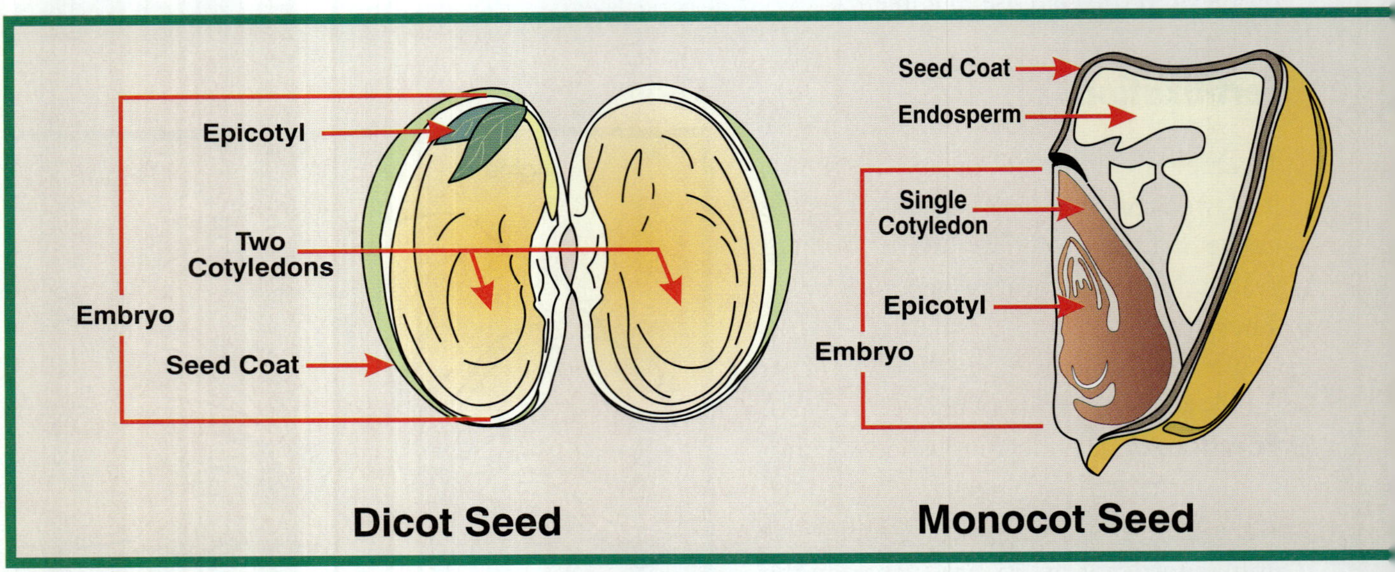

8–10. Parts of a seed.

The ***embryo*** is held in a dormant or resting phase inside the seed. It has a stem, root, and one or two seed leaves called ***cotyledons***. Embryos of monocotyledon (monocot) plants have one cotyledon, while those of dicotyledon (dicot) plants have two cotyledons. Monocot plants store the bulk of their energy in the endosperm. Dicots store their food in the two cotyledons. When germination begins, the embryo draws energy from the stored food to emerge from the soil and to develop leaves. Once the leaves have developed, the seedling can manufacture its own food.

8–11. The cotyledons of dicots provide food for the developing seedling.

Seed Germination

Seeds are designed to wait for favorable conditions to begin growth. The wait for favorable conditions may take many years. It is fascinating to think about how seeds can lay dormant for many years and suddenly spring to life when given the right conditions for survival.

One mechanism some seeds have to help ensure survival is called stratification. ***Stratification*** is the process whereby a seed must go through a period of cold temperatures before it will germinate. This mechanism makes sense. Without this waiting period, a seed could germinate during a warm spell in December and die with the arrival of freezing temperatures. Lilies, roses, and pansies are three examples of plants that require a stratification period to germinate.

Another dormancy mechanism is ***scarification***, or the breaking down of the seed coat. Some seeds have very hard, thick seed coats that provide excellent protection for the seed. However, the seed coat prevents the absorption of water and germination under normal conditions. In nature, these seeds must pass through the acid stomach of an animal to wear down the seed coat or lay in the soil where microorganisms can eat away the seed coat before the seeds can germinate. Some examples of plant seeds that require scarification are geranium, lupine, and canna.

CONDITIONS FOR SEED GERMINATION

What are the right conditions for seed germination? Water is necessary for seeds to germinate. At the same time, seeds need oxygen to germinate. Seeds need optimum temperatures. They may also have a light requirement. The exact need regarding each of these factors varies with the plant species.

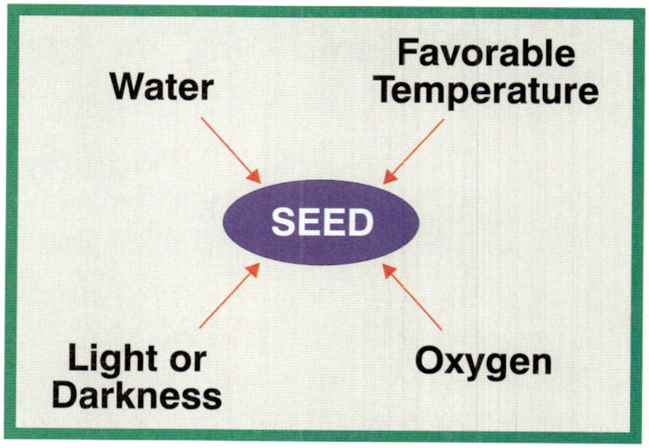

8–12. Seeds germinate when given the right conditions.

Also, it should be noted that if any one of these environmental factors is not favorable, the seed may not germinate.

The first step in the germination process is the absorption of water, which triggers a number of activities within the seed. The seed swells with moisture. The embryo changes from a dormant state to an actively growing plant. Respiration occurs. Proteins are synthesized. Gibberellins stimulate the production of enzymes. The enzymes, in turn, convert starches stored in the endosperm or the cotyledons to sugar. The embryo breaks through the seed coat and develops a primary root. Then, the stem of the embryo sprouts upward. After germination, the seedling needs light, water, and minerals to begin producing its own food.

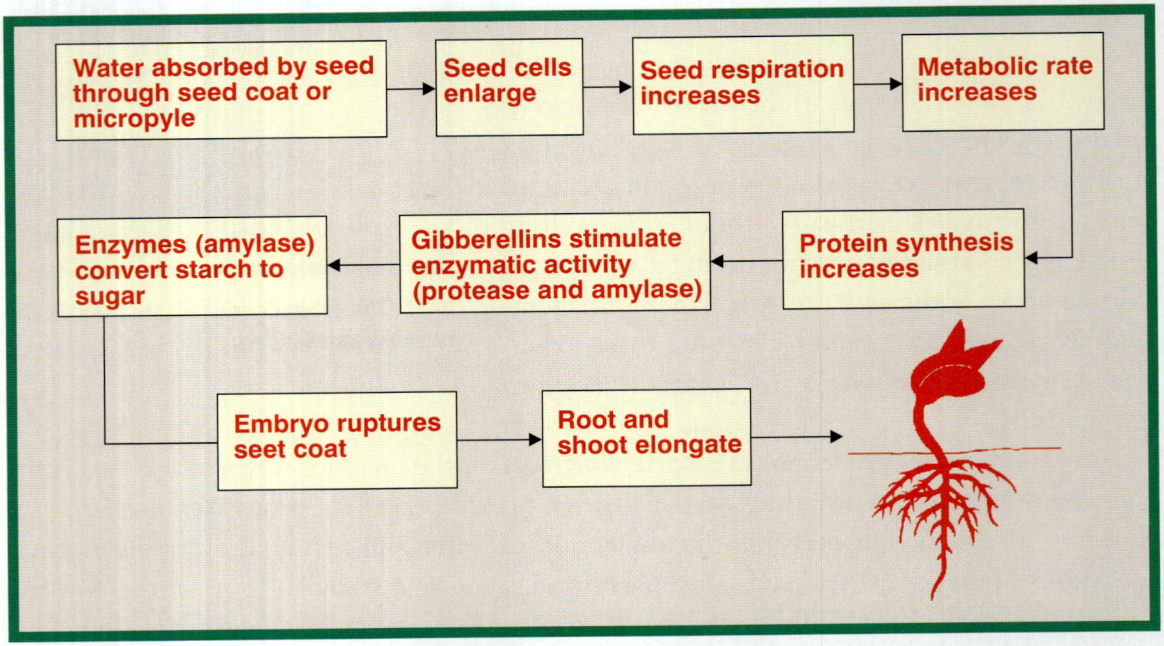

8–13. Steps of the seed germination process.

Seed Quality and Storage

The quality of seed used is very important to greenhouse growers. Seed quality refers to both viability and vigor. ***Viability*** is the ability of seeds to germinate under optimum conditions. ***Vigor*** is the ability of seeds to germinate under different conditions and still produce healthy seedlings. High quality seeds that produce usable seedlings are important to growers. Low quality seeds that fail to germinate are financial losses. In addition, they take space in the greenhouse and waste labor.

8–14. Proper storage maintains seed viability and vigor. At Pan American Seed, seeds are stored under cool, dry conditions. (Courtesy, Ball Horticultural Co.)

Store seeds in a cool dry location to maintain seed vigor and viability. Storage with a temperature of 40°F and a relative humidity between 20 and 40 percent is acceptable. Small growers often place their seed in sealed containers in a frost-free refrigerator. Growers for large greenhouse operations store their seed in coolers that deliver optimum storage conditions. No matter how they are stored, the quality of the seed will decrease over time.

A high percentage of seeds capable of producing usable seedlings is desired. Seed companies run tests to determine what percentage of the

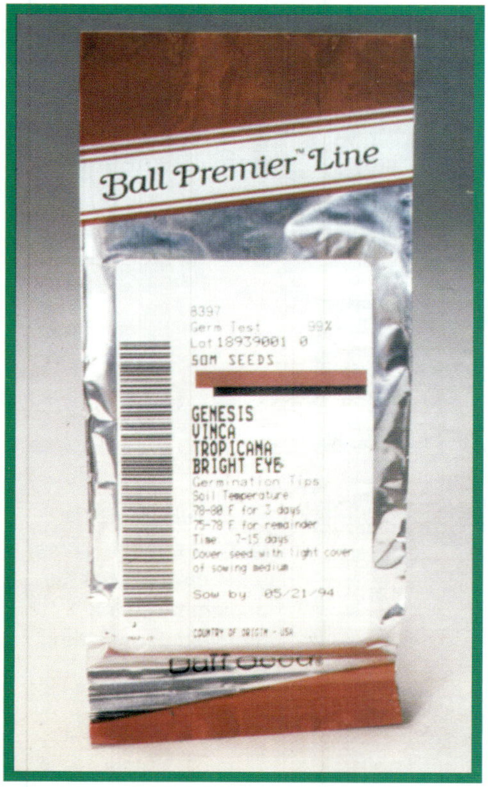

8–15. The germination rate and the year for which the seed is packaged are printed on the packet. (Courtesy, Ball Horticultural Co.)

8–16. Seed is often pelleted to ease handling. (Courtesy, Ball Horticultural Co.)

seeds will germinate. That information must be printed on the seed packet. The date for which the seed is packaged also appears on the package.

Seeds may be altered or enhanced to improve performance or handling. They may be de-winged, de-fuzzed, or de-tailed for ease of handling. They might also be pelleted, coated, scarified, or primed. However, enhancing the seeds in these ways shortens the period of time the seed can be stored.

Sowing Techniques

Use of proper techniques in sowing seed improves germination results. To get good results, the grower first needs to recognize optimum germination conditions for the particular seed being planted. Does the seed need to be stratified or scarified? Does the seed need to be exposed to light or covered? How moist should the medium be and at what temperature should the medium be maintained? These are some questions that should be addressed.

The seed treatment for most floriculture crops is similar. In general, the growing medium temperature for seed germination is around 75°F. Large operations have germination rooms that provide the optimum temperatures. On a smaller scale, the seed flats can be kept warm with the use of germinating pads or systems that pump warm water through tubing under the seed flats.

Most seeds germinate in darkness and are planted at a depth three times the depth of the seed. Planting seeds too deep can make it difficult for the embryo to emerge. If planted too shallow, the moisture level may not be optimum or the seed may be exposed to light when it needs darkness.

Maintaining a uniform moisture level in the medium is very important. If the medium dries during the germination process, the seeds will likely die. Moisture levels can be maintained by covering the seed flats with plastic or glass.

8–17. Warm water is pumped through tubing under flats to keep the medium warm for improved germination.

8–18. Automated seeders sow seeds directly into plug trays. (Courtesy, *Greenhouse Product News*)

Automation has changed seeding practices in many greenhouse operations. Machines costing thousands of dollars place seed directly in flats. However, hand seeding is performed in smaller operations. Some practices for successful seed germination follow:

- Start with clean containers with drainage holes.
- Use a clean, fine-textured medium prepared for seed germination.
- Level the medium in the seed flat, while firming it slightly. Moisten the medium prior to sowing the seed.
- Use fresh, viable seed. Do not open the seed packet until it is time to sow.
- Sow the seed in rows to reduce the spread of disease, if it is introduced.
- Sow the seeds at a proper depth. If recommended, cover the seed with medium.
- Label the flat with the seed variety planted and the date sown.
- Mist the flat to ensure thorough wetting of the medium and to improve contact between the seed and medium. Avoid watering practices that create a heavy flow, as it can wash medium and seeds out of position.
- Maintain proper medium temperature. Monitor the medium temperature, not the air temperature.
- Cover seed flats with poly or glass to maintain the moisture level and humidity. Black poly might be used to ensure darkness but must be removed once the seeds start to germinate.

- Transplant the seedlings when they have developed their first set of true leaves. Do this by carefully lifting the seedlings from the medium. Avoid tearing the roots. Also, handle the seedlings by their leaves or cotyledons so as not to damage the easily bruised stems.

8–19. Techniques for hand seeding include: (a) add medium and level, (b) moisten medium, (c) make trenches for seed, (d) sow seed, (e) cover seed as required, and (f) label flat.

ASEXUAL PROPAGATION

Asexual propagation is plant reproduction using the vegetative parts of a plant. Leaves, stems, or roots might be used. Unlike sexual propagation, asexual propagation results in offspring that are the genetic duplicates, or ***clones***, of the parent plant.

There are certain advantages associated with asexual reproduction:

8–20. The yellow chrysanthemums in this photograph were propagated asexually and are therefore genetically identical.

Recombination of genes that would occur in sexual reproduction does not occur. Therefore, plants with outstanding characteristics can be produced without the risk of losing the desired characteristics, such as flower color or leaf variegation. Some plants are difficult to reproduce sexually. They may produce few seeds or the seeds they produce may result in low germination rates. Huge numbers of genetically identical plants (clones) can be produced. Plants can be grown that are free of diseases. Mature plants are obtained more rapidly as compared to seed.

Methods

There are many methods of propagating plants asexually. Some plants are easily propagated in a number of ways. Other plants respond best to a particular method. Cuttings, grafting, layering, separation, division, and tissue culture are types of asexual propagation methods.

Many plant varieties and cultivars are patented, much like an invention is patented. If they are propagated, a royalty is paid to the owner of the patent. Failure to do so is against the law. Some companies specialize in providing rooted cuttings and stock plants

8–21. These rooted chrysanthemum cuttings are newly potted.

8–22. Cuttings are taken from stock plants.

to growers. Chrysanthemums are a good example of a floriculture crop propagated by specialists. Most greenhouse growers buy rooted chrysanthemum cuttings and pot them up. The royalties are factored into the cost of the cuttings.

Herbaceous Cuttings

One of the most common and simplest methods of asexual propagation is that of ***cuttings***. Cuttings may be made from portions of stems, leaves, or roots. Most greenhouse crops are propagated by herbaceous or soft stem cuttings. Growers often keep ***stock plants*** or buy new stock plants each year from which stem cuttings are removed. The apical meristems of stock plants are pinched to encourage branching. Other types of herbaceous cuttings include leaf cuttings, leaf-petiole cuttings, and leaf-bud cuttings.

Once taken, cuttings are given optimal environmental conditions to promote the regeneration of the missing plant organ or organs. Environmental conditions critical for successful rooting include proper temperature, high humidity, and sufficient light. The rooting media must be free of disease organisms. It should also have good water-holding ability and good aeration.

Rooting is a complex physiological process. The time it takes for the cuttings to root depends on many factors. The plant species or variety, the age of the plant, the type and location of the cutting, the absence or presence of leaves, and the nutritional status of the plant influence rooting.

8–23. The most widely used asexual propagation method for floriculture crops is stem cuttings.

Stem Cuttings

Propagation by ***stem cuttings*** is the preferred method for poinsettias, chrysanthemums, zonal geraniums, kalanchoe, carnations, and many foliage plants. The cuttings, 2 to 5 inches in length, are taken in the morning when the tissues are turgid or full of water. A sharp knife or razor is used for the cuts. It is also advisable to disinfect the knife between stock plants to avoid spreading disease. Any flowers on the stem should be removed.

8–24. These poinsettia stem cuttings were stuck in rock wool rooting cubes.

8–25. Sticking cuttings directly into finish pots.

The distal end of the cutting, or the end closest to the root system, is usually dipped in a rooting hormone to hasten the formation of adventitious roots. The cuttings are then placed in rooting cubes or directly into a medium. This practice of placing the cuttings is known as ***sticking cuttings***.

After the cuttings are stuck, they are placed under an intermit-

8–26. After sticking, cuttings are placed under an intermittent mist system and shaded.

tent mist system. The mist is applied between dawn and sunset to reduce water loss from transpiration. They are misted regularly until the roots form and can absorb moisture for the plant. The propagation area is often shaded to reduce stress from the sun. Bottom heat is also provided to maintain a medium temperature between 75°F and 80°F.

More and more growers use the practice of direct sticking cuttings into the finish container to save labor involved in

8–27. In the home, plastic bags are placed over the container to maintain high humidity around the cuttings.

8–28. Dumbcane or Dieffenbachia can be propagated by placing sections of the stem horizontally or vertically in the medium.

transplanting. However, the grower must have enough bench space under mist and a means to keep the medium warm.

Leaf Cuttings

A relatively small number of plants have the ability to produce plantlets on their leaves. Entire leaves or portions of a leaf are removed from the parent plant for use as ***leaf cuttings***. Plants propagated by this method include African violet (Saintpaulia ionantha), Mother-in-law's tongue (*Sansevieria trifasciata)*, Cape primrose (*Streptocarpus x hybridus)*, and Rex begonia.

Propagation of Mother-in-law's tongue, Cape primrose, and Rex begonia involves the placement of a leaf blade or a portion of a leaf blade in a propagation medium. Healthy leaves that have just reached maturity should be used. Those leaves are at a stage when food production and the capacity to produce new plantlets are high.

African violets are easily propagated by ***leaf-petiole cuttings***. The leaf blade and the petiole are taken from

8–29. Sansevieria is propagated by portions of a leaf.

8–30. The leaf blade of the Rex begonia laid on medium will produce plants where leaf veins have been cut.

8–31. African violet propagation is done by leaf-petiole cuttings.

8–32. English ivy is propagated by leaf-bud cuttings.

8–33. In one day, this Floridian can stick 10,000 leaf-bud cuttings directly into finish pots.

the parent plant. Under proper conditions, a cluster of plantlets develop where the petiole was cut. When the plantlets have grown large enough to handle, they are separated and potted in individual containers.

A few plants, including the piggyback plant (*Tolmiea menziesii*), kalanchoe, and the strawberry geranium, produce foliar embryos. Through a complex process, cells in small areas of a leaf develop into plantlets. The plantlets can be removed and planted.

Leaf-bud Cuttings

Many plants cannot be propagated by leaves alone. One option is to take leaf-bud cuttings. ***Leaf-bud cuttings*** are composed of the leaf blade, the petiole, a bud at the base of the petiole, and a portion of the stem. The leaf provides the energy for the development of roots. The roots sprout from the stem and are often concentrated at the node of the stem. The bud develops into the stem of the new plant.

Grafting

A method of asexual reproduction common with woody plants is grafting. ***Grafting*** is the process in which the stem of one plant is made to grow on the roots of another plant. The portion of the graft that is to become the stem is the ***scion***. The lower portion of the plant that includes the root system is called the ***rootstock*** or the understock. ***Budding*** is a form of grafting in which the scion consists of a bud.

Roses are grafted. Usually, a vigorous, cold-tolerant species of rose is selected for the rootstock. Vigorous rootstocks provide an abundance of water and minerals to the scion. A desired rose hybrid that might not have a strong root system is grafted to the rootstock.

Grafting may involve placing individual buds into a stem or a stem being placed onto another stem. In either case, it is important that the cambium wood of both the scion and the rootstock line up. Once placed together, the union should be protected from moisture loss. Also, the scion and the rootstock materials must be capable of growing together for a successful graft.

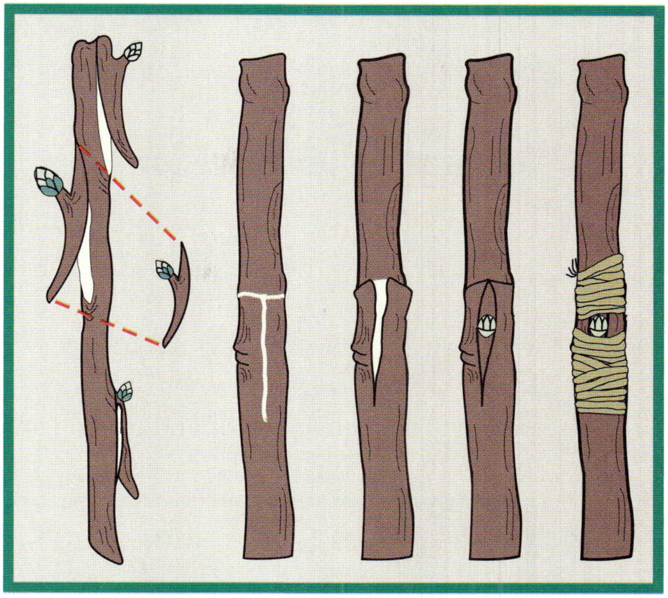

8–34. Budding is a form of grafting.

Layering

Layering is a method of asexual reproduction whereby roots form on a stem while the stem is still attached to the parent plant. The advantage to layering is the parent plant provides the plant-to-be with water and minerals until it produces its own roots. This method is

8–35. Air layering is performed with some foliage plants.

slow as compared to other methods. A form of layering called ***air layering*** is used with foliage plants including *Dieffenbachia*, *Ficus*, and *Dracaena.*

Separation and Division

Propagation of floriculture crops can be done by separating or dividing. This is a common method used with perennials and foliage plants. Some plants produce vegetative plant structures that can be removed intact from the parent plant. Removal and planting of these vegetative structures is ***separation***. With ***division***, the plant roots or the entire plant may be cut into sections to make two or more plants from the original plant. Daylilies can be divided by digging a plant and cutting it into smaller portions.

8–36. Separation involves the removal of whole vegetative plant structures from the parent plant.

There are several methods used to propagate bulbs. Some species of lilies produce ***bulbils*** or tiny above-ground bulbs in the axils of their leaves. These can be removed and planted. Lilies also may produce tiny bulbs below the ground called ***bulblets***. Some lilies and fritillaries can be propagated by removing bulb scales and placing them in moist medium. In time, they root and produce bulblets that can be separated and planted.

Tulips and *Narcissus* reproduce by natural division. Bulbs are produced off the main bulb. These are separated and planted. Hyacinths are very slow to reproduce by natural division. They can be encouraged to produce bulblets by scooping or scoring. ***Scooping*** involves the removal of the basal plate of the bulb and the base of all of the bulb scales. Placed upside down

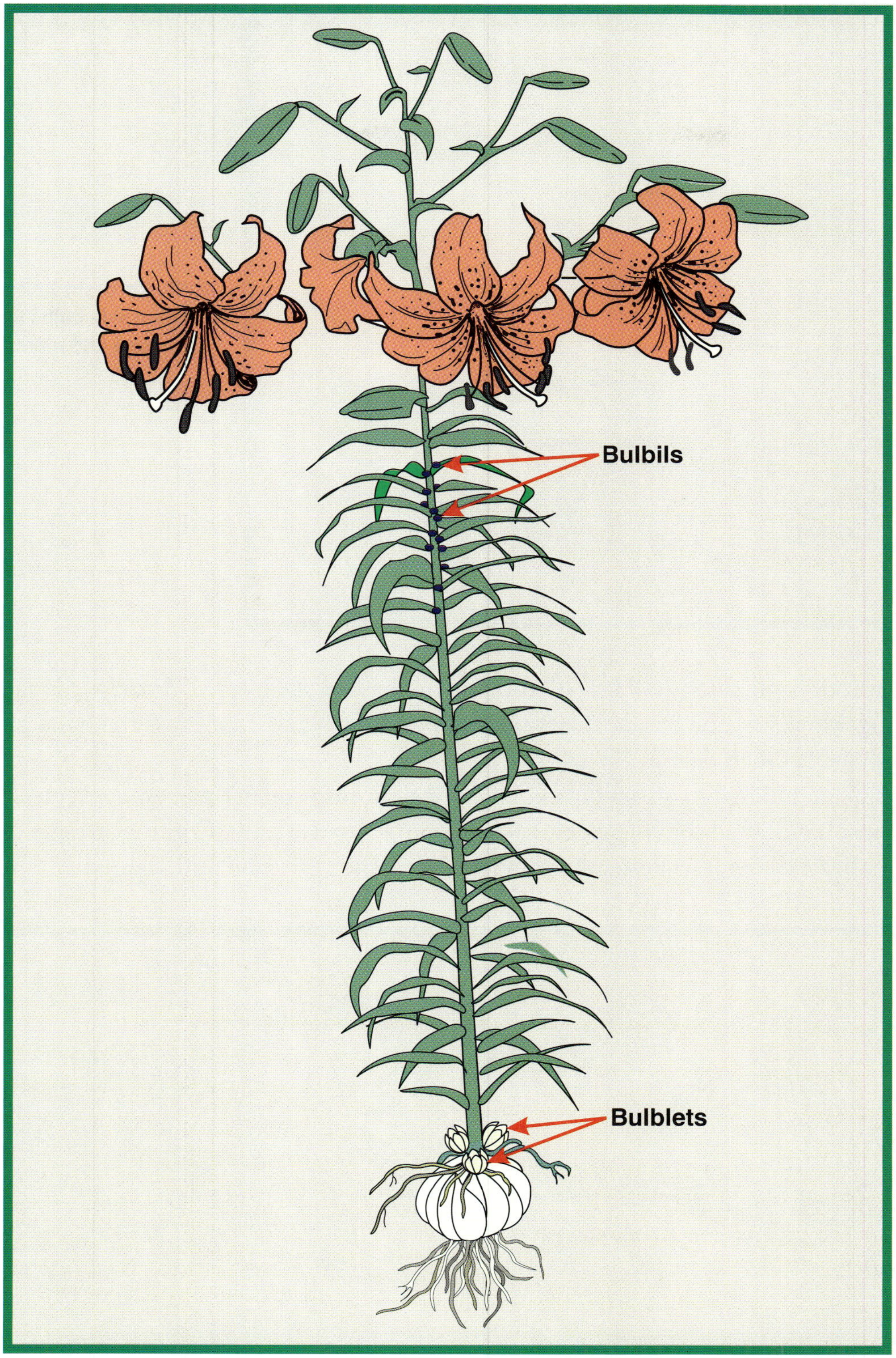

8–37. Lilies may produce bulblets below the ground or bulbils in the axils of the leaves. The bulbils can be removed and planted.

8–38. Narcissus bulbs produce side bulbs that are separated and planted.

in a warm dry cabinet, a bulblet will form at the base of each scale. ***Scoring*** is similar to scooping. However, the basal plate is not removed. Two cuts that cross the basal plate are made about 1/4 inch deep.

Corms, including crocus and gladiola, can be cut into smaller pieces. Each piece of the corm must have a bud that is capable of developing into the stem. Corms also develop small corms called ***cormels***. These miniature corms can be separated and planted.

8–39. Scoring, done with hyacinths, involves crossing cuts into the base of the bulb.

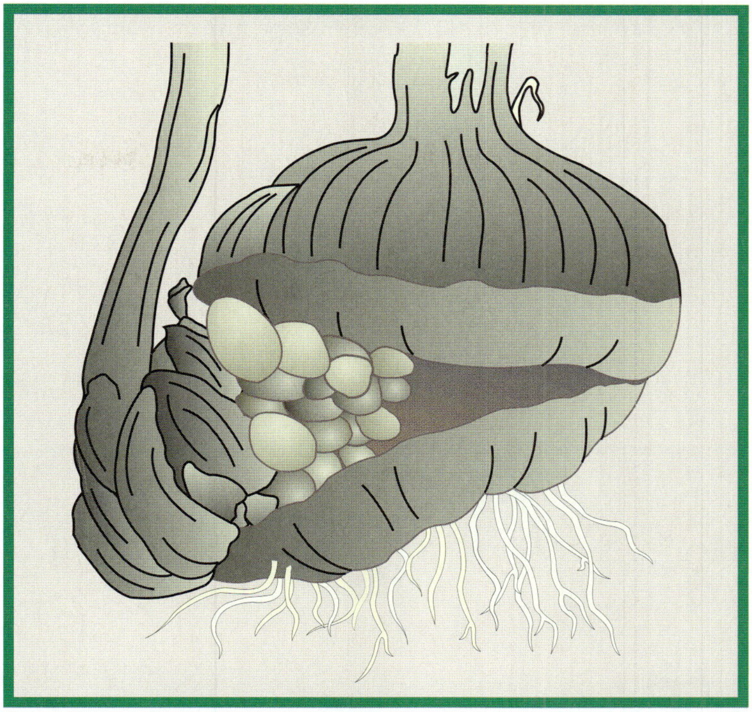

8–40. Gladiolas produce cormels that are separated and planted.

Tissue Culture

A very technical method of asexual propagation is ***tissue culture***. Tissue culture involves the culture or growing of small pieces of plant tissue. It is performed on an artificial medium under sterile conditions. Foliage plants, pot plants, and cut flowers are propagated by tissue culture methods.

There are several advantages to tissue culture over other methods of propagation:

1. Large numbers of plants can be produced from a single plant in a relatively small space in a short period of time.
2. Systemic diseases can be eliminated by propagating the quickly dividing cells of the shoot tip.
3. The ability to produce plants with identical flower color is an asset to the cut flower industry.
4. Tissue culture results in excellent basal branching of foliage plants.
5. Horticultural cultivars can be improved by selecting plants that vary slightly from the mother plant. Examples are leaf shape, disease resistance, growth habit, and flower color.

6. Tissue culture techniques permit the growth of genetically engineered plant cells.

Tissue culture must be done under aseptic or sterile conditions. Tissue culture labs are designed to provide a clean environment. Technicians scrub much like a surgeon does before surgery. Media, tools, and bottles or jars are autoclaved. Autoclaving involves the heating of the materials to 245°F for 15 minutes to kill all bacteria and fungi. Also, to lessen the chance of contamination, work with the cultures is done under laminar air-flow hoods. Laminar air-flow hoods filter bacteria and fungal spores from the air.

8–41. This lab technician prepares medium before it is autoclaved.

8–42. Tissue culture is performed under laminar air-flow hoods.

Table 8–1. Some Floriculture Crops Propagated by Tissue Culture

Foliage Plants	Potted Flowering Plants	Cut Flowers
Alocasia	African violet	Alstroemeria
Anthurium	Aloe vera	Asters
Banana	Gerbera daisy	Delphinium
Calathea	Lily	Gerbera daisy
Cordyline	Orchid	Gysophila
Dieffenbachia	Rose	Lily
DracaenaFern	Rex begonia	OrchidRose
Ficus		Statice
Philodendron		
Spathiphyllum		
Syngonium		

TISSUE CULTURE PROCESS

The tissue culture propagation process can be defined in four main stages. The first three stages must take place under sterile conditions to reduce the chance of contamination from bacteria or fungi.

In the first stage of tissue culture, small pieces of plant material, called ***explants***, are carefully removed from the parent plant. Explants are obtained from the growing tip of a desired plant. The explants are cleaned of bacteria or fungal spores. They are then placed on an agar media in glass bottles or test tubes. The agar media is a gel that contains water, sugars, nutrients, and plant hormones to support and promote plant growth.

In stage two, the cells of the explants multiply in one of two ways. They may form a ***callous***, which is a group of cells with no particular function. Given the right hormones in the media, these callous cells can develop into a normal plant.

8–43. Plantlets are divided and encouraged to multiply in stage two.

The other possibility for stage two involves the rapid multiplication of plantlets. Cytokinins placed in the media encourage an increase in the number of buds on the explants, usually six to eight per shoot. Each bud is capable of

8–44. A single explant can produce millions of plantlets in a year by continually cycling plantlets to new containers.

becoming a plant and producing more buds. Branching occurs as these buds develop into shoots or **plantlets**. These plantlets are divided and transferred to test tubes or jars. New plantlets are continually cycled to new containers. In this way, a single explant can produce millions of plantlets in a year.

When the plantlets have developed, they are ready for the third stage, the formation of roots. Shoots are transplanted to another medium containing auxins that induce the growth of roots. The plantlets are also given higher light intensity in preparation for stage four.

In stage four, the plantlets are removed from the glass container. They are divided, planted in a sterile growing medium, and placed in a greenhouse. The humidity in the greenhouse is much lower than in glass containers. Therefore, care must be taken during this transition to acclimatize the plants to their new environment.

The use of tissue culture is growing rapidly in the floriculture industry. It is used to a large extent with herbaceous perennial plants including daylilies, Hostas, chrysanthemums, and orchids. The importance of tissue culturing in plant breeding will become more significant as the field of genetic engineering expands.

8–45. In stage four, plantlets are removed from the glass container and planted into sterile growing medium.

8–46. Plantlets are misted to acclimatize them to greenhouse conditions.

GENETIC ENGINEERING

Traditional methods of plant breeding through pollination have left plant breeders dependent on luck. They have had no control over chromosomal division in meiosis. Whether a desired trait would be expressed in a hybrid was in question. Even the benefits of polyploidy have been left to chance. Another disadvantage of traditional breeding is the available genetic material is limited to that found in a certain species. Only a chrysanthemum can be crossed with a chrysanthemum.

However, plant breeding practices are being dramatically changed with new discoveries in biotechnology and genetic engineering. Recent discoveries in science have enabled scientists to select and move genetic material from one plant to another. Genes can even be transferred from one species to another. This process, called **genetic engineering**, holds great potential for improving floriculture crops. Desired results can be obtained much more quickly than with traditional breeding methods. There is also greater control over what characteristics will be expressed in the offspring.

Genetic engineering has already been applied to vegetable crops and grains. Genetically engineered squash that is resistant to viral infection has been developed. Soybeans that are resistant to herbicides have been developed. Genetically engineered corn and cotton have improved defenses against insect pests.

Imagine what could be done in the floriculture industry. The aesthetic value of floriculture crops could be improved. A true blue rose is possible. Cut flowers that have a vase life of a month could be a reality. Crops could be engineered to be resistant to soil-borne diseases, cutting costs and labor involved in disease control. Poinsettias could be engineered to resist white fly infestations. The possibilities are endless.

REVIEWING

MAIN IDEAS

Plants are propagated in two ways. They may be propagated sexually or they may be propagated asexually. Sexual propagation involves flowers and seeds. Pollination of a flower occurs when pollen is transferred to a stigma. Fertilization takes place when sperm carried in the pollen fuses with the egg held in the ovule. The result of sexual reproduction is a seed. The offspring is a hybrid of the parent plants. The significance of sexual propagation is the combining of genetic material from two parents.

Seeds produced through sexual reproduction contain a living embryo and food to help it begin growth. Quality seed is viable and has vigor. Germination or the start of growth happens when the seed receives the right amounts of water, oxygen, temperature, and possibly light. Growers maintain optimum conditions for seed germination.

Asexual propagation involves the vegetative parts of a plant; leaves, stems, or roots. The offspring resulting from asexual propagation are genetically identical or clones of the parent plant. There are many methods of asexual propagation. Methods of asexual propagation include cuttings, grafting, layering, separation, division, and tissue culture. The most widely used method in the floriculture industry is stem cuttings. A newer, highly technical method of asexual propagation is tissue culture. Tissue culture consists of placing explants on an agar gel under sterile conditions.

Recent research in biotechnology has given scientists and plant breeders the ability to improve plants through genetic engineering. Genetic material in the cell of an organism can be removed and placed into another organism through genetic engineering. This technique has tremendous potential for improving floriculture crops.

QUESTIONS

Answer the following questions using complete sentences and correct spelling.

1. What is the major difference between sexual and asexual propagation?
2. How does pollination occur?
3. What is the advantage of hybridization?
4. Why are polyploid plants valued?
5. How does fertilization occur?
6. What are the main parts of a seed and their function?
7. What factors are involved in seed germination?
8. What are stratification and scarification?
9. What are the advantages of asexual propagation?
10. How are stem cuttings made?
11. How is tissue culture performed?
12. What are the potential advantages to genetic engineering?

EVALUATING

Match the term with the correct definition. Write the letter of the term on the line provided.

a. tissue culture
b. explant
c. pollination
d. fertilization
e. polyploid
f. viability
g. stratification
h. germination
i. leaf-bud cutting
j. stock plants
k. grafting
l. asexual propagation

_____ 1. The ability of seeds to germinate under optimum conditions.

_____ 2. Plants from which cuttings are removed.

__f__ 3. Beginning of growth from a seed.

__g__ 4. Transfer of pollen from the male part of a flower to the stigma.

__J__ 5. When sperm carried in the pollen fuses with the egg held in the ovule.

__k__ 6. Mechanism that requires a seed to go through a cold period before it will germinate.

__h__ 7. Plant reproduction using vegetative parts of a plant.

__B__ 8. Cutting composed of the leaf blade, the petiole, a bud at the base of the petiole, and a portion of the stem.

__C__ 9. Asexual method of reproduction in which the scion is fused with the rootstock.

__A__ 10. Small portion of a plant used in tissue culture.

__e__ 11. Plants that have more than the normal double set of chromosomes.

_____ 12. Highly technical method of asexual reproduction involving the culture or growing of small pieces of plant tissue.

EXPLORING

1. Obtain different types of cuttings from various houseplants. Place them in a growing medium. Build a tent made of clear poly over the container to maintain high humidity. Set the container by a window. Observe the cuttings for several weeks to determine if adventitious roots begin to grow.

2. Sow bedding plant seeds following the guidelines in the chapter. Schedule the sowing so the seedlings will be ready for planting outside in the spring.

3. Use the World Wide Web to learn more about applications of plant biotechnology in the floriculture industry.

CHAPTER 9

Cultural Practices: Light, Temperature, Air, Water

OBJECTIVES

This chapter is about the effects of light, temperature, air, and water on plant growth. It has the following objectives:

1. Explain how light influences greenhouse crop production.
2. Discuss the plant responses to light duration.
3. Express the theory behind DIF.
4. Describe the effects of temperature on greenhouse crop production.
5. Explain how air quality influences plant growth.
6. Explain the importance of water quality.
7. Describe recommended watering practices.

TERMS

capillary mats
cultural practices
day-neutral plants
drip irrigation
ebb and flood method
electromagnetic spectrum
far-red light
foot candle
hand watering
high-intensity discharge (HID) lighting system
intermittent mist system
irrigation
irrigation booms
long-day plants
overhead sprinklers
photoperiodism
saucers
short-day plants
spaghetti tubing
thermoperiodism
turgidity
vernalization
wilting

9–1. Growers understand the needs of plants.

HAVE you ever heard someone joke about having a brown thumb instead of a green thumb? People who claim to have a brown thumb say that every plant they have had has died under their care. On the other hand, a person with a green thumb appears to grow plants without effort. In truth, the ability to grow healthy plants is not automatic. Care of plants can be learned.

Plants have basic needs. If those needs are met, the plants grow and thrive. If they are not met, they decline and die. Part of the success with growing healthy plants is in knowing what the plant needs. Another part is providing those needs at the appropriate times. Top growers have learned to understand the needs of plants. They draw from knowledge gained in the classroom, from talking with other growers, and by applying their skills in the greenhouse.

CULTURE OF GREENHOUSE CROPS

Most of us learned at an early age that plants need water and light to grow. Later we might have learned that temperature and air are also important. Experienced growers know how to control these environmental factors to produce floriculture crops. The control or management of light, temperature, air, and water to promote healthy growth is often referred to as ***cultural practices***. Understanding basic cultural requirements of greenhouse crops is of great importance.

9–2. The growth and development of floriculture crops is dependent on light energy.

LIGHT

The growth and development of floriculture crops is dependent on light energy. Light energy enables plants to make food through the process of photosynthesis. Light also influences plant growth in other ways. There are three key aspects of light affecting plant growth. These are the color of the light, the duration of the light, and the intensity of the light.

Light Color

Visible light or the light that we see is a small segment of all the radiant energy given off by the sun. X-rays, gamma rays, ultraviolet rays, microwaves, and radio waves are some others. The wavelengths of the different rays are measured in nanometers. Based on their wavelengths, the rays have been placed on an ***electromagnetic spectrum***.

Sunlight contains a complete blend of visible colors including red, orange, yellow, green, blue, and violet. The rays of visible light fall between 380 and 780 nanometers. Each color has a different range of wavelengths. We can see the individual colors when the light passes through a prism. The prism separates the colors. This phenomenon of separating the colors of visible light occurs naturally with rainbows.

Light is composed of particles called photons. They carry with them energy. The energy level of the photons increases as the wavelengths become shorter. For example, blue light photons

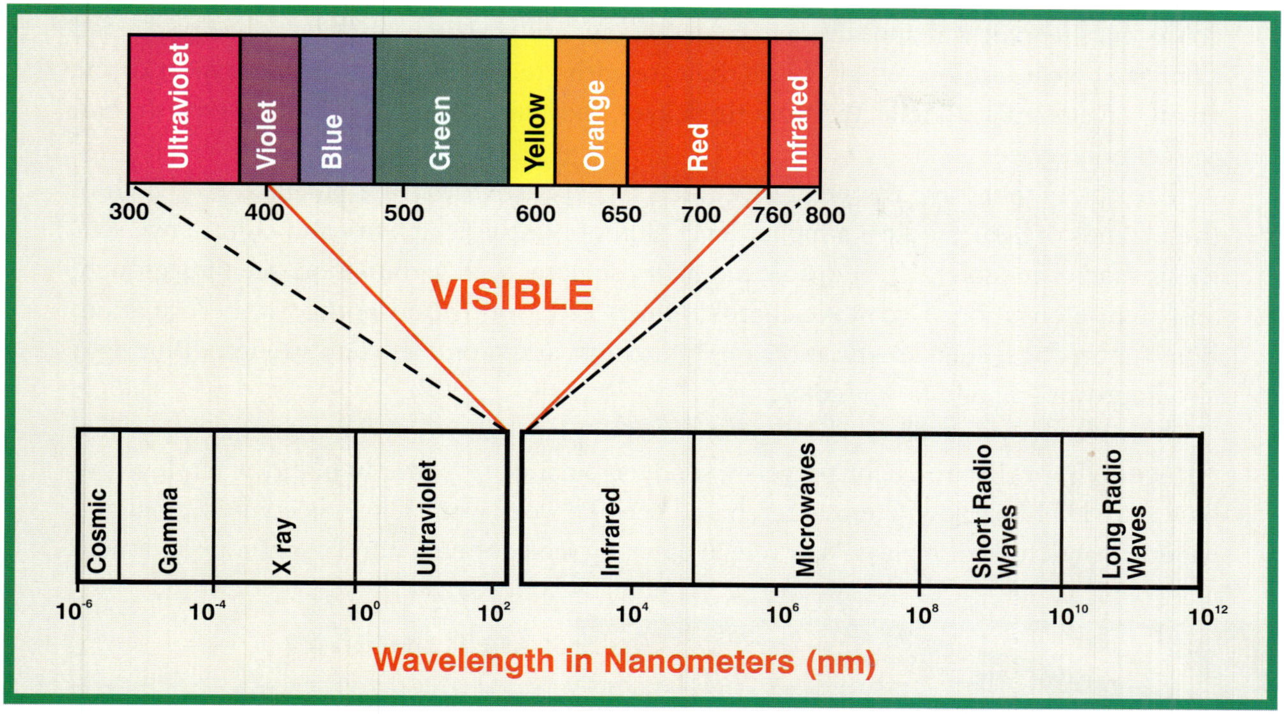

9–3. The electromagnetic spectrum.

have a shorter wavelength than red light photons, and therefore have a higher energy level. The energy carried by the photons is absorbed by objects and produces heat. Plants, through photosynthesis, are capable of converting light energy to stored chemical energy.

The visible rays of light are absorbed or reflected by objects. Objects that appear black absorb all the colors of visible light. Therefore, no colors are reflected into our eyes. White objects, on the other hand, reflect all the colors of visible light. If an object absorbs all the

9–4. The flowers and leaves of these gerbera daisies reflect colors of visible light into our eyes. (Courtesy, Society of American Florists)

colors but green, it will appear green to us because green wavelengths have been reflected from the object into our eyes. This is the case with green leaf plants. Because most of the green wavelengths are reflected from the leaves, they have less effect on plant growth than the other colors.

The colors that have the greatest influence on plant growth are blue and red. Both of these bands of light produce a higher level of photosynthetic activity and chlorophyll synthesis than other colors. In addition, red light promotes seed germination, seedling growth, and stem elongation. Red wavelengths influence flowering and anthocyanin formation. Blue light reduces stem length, increases branching, and promotes stem strength. Blue light also improves leaf and flower color.

A wavelength that plays an important role in plant growth and development is called ***far-red light***. Far-red light triggers a shade avoidance response in plants when levels of blue or red light are low. Stems stretch and become weak while leaves become thinner and wider. Far-red light also plays a key role in breaking seed dormancy and in photoperiodism.

9–5. Kalanchoes must receive short-day treatment to flower.

Light Duration

Plants are responsive to the length of time they are exposed to light. A mechanism within plants detects day lengths. The length of the days or the length of the light period is known to influence different phases of plant growth. Some of these phases are seed germination, enlargement of leaves, development of buds, and flowering. A plant's response to light duration is called ***photoperiodism***.

Different plants respond differently to light duration. Chrysanthemums, kalanchoe, holiday cacti, and poinsettias are plants that will begin to flower naturally in the fall when the day lengths get shorter. They are said to be ***short-day plants***. Carnations, Shasta daisies, and coneflower plants flower as days lengthen in the summer. They are termed ***long-day plants***. A third group of plants is unaffected by day length and are classified as ***day-neutral plants***. Day-neutral plants include many foliage plants, African violets, and tomatoes.

Growers control the length of days in a greenhouse to bring on one of two plant responses. The plants will either be kept in a vegetative stage of growth or encouraged to

9–6. This photograph shows black cloth curtains in a partial closed position. (Courtesy, United Greenhouse Systems, Inc.)

flower. The key is not so much the length of day as the length of uninterrupted darkness. For example, poinsettias will begin to initiate flower buds when the length of uninterrupted darkness is about 11 hours and 50 minutes. Poinsettias respond to nights shorter than 11:50 by growing vegetatively. If the darkness is interrupted by someone turning on the lights, flowering will be delayed. That is because the poinsettia thinks it is in a long-day situation.

Short-day plants can be made to flower when the days are long. The plants are covered with black cloth to extend the period of darkness. Thinking it is in a short-day situation, the plant begins to initiate flower buds. This is why you can find flowering chrysanthemums in stores at any time of the year. To keep a short-day plant in a vegetative stage of growth, they are lighted at night to simulate long days. Managing the day length is critical for getting plants to flower at a time of year when they normally would not flower.

9–7. Some crops are lighted at night to simulate long days.

Light Intensity

The third effect light has on plants involves intensity or brightness. Intensity of light depends largely on the angle of the sun, clouds, and dust in the atmosphere. Light intensity is greater in the summer months when the sun is higher in the sky. Where artificial light is provided, the source of light and the distance from the light source to the plants are important.

Light intensity is measured in foot candles. A ***foot candle*** is the amount of light distributed by a single candle one foot away. A sunny, summer day provides about 10,000 foot candles of light.

Different plants have different requirements in terms of intensity of light. The major floriculture crops perform the best when they receive full sunlight. The high light intensity elevates the rate of photosynthesis and the plant is able to produce more food. On the other hand, plants like impatiens, African violets, ferns, and many foliage plants prefer lower light levels. In fact, most foliage plants have been selected because of their ability to tolerate low levels of light. Exposure to full sun can damage these crops.

Light intensity influences plant development. Plants receiving the proper level of light will be compact and have good leaf color. Symptoms of lower than optimum light levels include a slower growth rate, thin leaves, small flowers, dull leaf and flower color, and etiolation or stretching. High light intensity can also cause problems. Plants receiving too high of light intensity might have pale leaves, flowers, and stems. Other symptoms include short rigid stems, small thick leaves, and small flowers. Excessive light intensity may even cause death of leaf tissue.

Reducing Light Intensity

Light intensity in the greenhouse can be reduced if necessary. The amount of light that is transmitted varies with the greenhouse covering material. So if a grower knows the crop to be grown prefers lower light, a covering can be

9–8. Thousands of hanging baskets create shade on the bedding plants below.

9–9. The glazing on the right side of this greenhouses has been whitewashed to reduce light intensity.

selected that blocks more light. If the greenhouse has a covering, such as glass, that provides high light transmission, it can be whitewashed. This is accomplished by mixing one part white latex paint to 15 to 20 parts water. It is then sprayed onto the outer surface. Whitewashing also reduces heat buildup in the greenhouse. Over a period of several months, the paint wears off. A third option is to shade the crops with curtains.

9–10. Curtains in this photograph have been drawn to reduce light intensity.

Supplemental Lighting

Supplemental lighting is valued in greenhouse operations located in low light regions of the country. It is particularly important during the months of November, December, and January. Also, artificial lighting during cloudy periods improves plant growth and vigor. One application involves the lighting of plants early in their development. Plugs and newly planted cuttings are given additional light to speed growth and to ensure compact healthy growth. Lighting is also used to promote a long-day response in some crops.

9–11. Supplemental lighting is used to get young plants off to a strong start. (Courtesy, P. L. Light Systems)

There are a number of light sources that can be used. However, none of them can provide the range of wavelengths given by the sun. Desirable features to consider when selecting a light source include high

9–12. Lighting systems, seen through the glass, can be used to supplement natural light.

efficiency, uniform light distribution, and limited shading. Until recently, incandescent and fluorescent lights were the only choices. The development of ***high-intensity discharge (HID) lighting systems*** have given growers more efficient and effective light sources.

Several sources can produce supplemental light. Among these are incandescent lights, fluorescent lights, metal halide lamps, low pressure sodium lamps, and high pressure sodium lamps.

Incandescent lights are considered inefficient, emitting most of the energy as heat. The wavelengths

9–13. This incandescent light turned on at night keeps these mums in a vegetative state.

9–14. Fluorescent lights are used in this tissue culture lab.

given off are primarily far-red. They are inexpensive and a good choice for photoperiod control.

Fluorescent lights are cooler and more efficient than incandescent lights. They also have a wider range of wavelengths and longer life. They are often used with incandescent lights to provide a more complete spectrum of light. Because of their large size and shading problems, they are used only in special situations.

Metal halide (MH) lamps display a spectrum of light similar to the sun's light. It is the best lamp for plant growth. However, they are not as efficient, long-lived, or cost-effective as other HID lamps.

9–15. High-pressure sodium lamps give off high levels of yellow, orange, and red wavelengths.

Low pressure sodium (LPS) lamps are the most efficient HID light source. They are expensive and shading becomes a problem due to their large size.

High pressure sodium (HPS) lamps emit a high level of red, yellow, and orange wavelengths. The combination of features, including their light output, efficiency, and long lives, makes them the best HID supplemental lighting system.

9–16. These narcissus, azalea, and Easter lilies have been placed in a cooler to slow their development.

TEMPERATURE

Temperature is an important factor governing plant growth. Important biochemical reactions, including photosynthesis and respiration, are affected by temperature. Temperature is a key factor in a number of plant responses. Life processes occur more quickly as temperature rises. They slow down as temperatures become cooler. This is because

9–17. Holiday cacti are thermoperiodic, meaning temperature plays a role in the initiation of flowers.

the enzymes that drive those reactions are sensitive to temperature. If the temperature becomes too warm or too cold, the enzymes are unable to carry out their functions.

9–18. Easter lilies must undergo vernalization to flower.

Temperatures can be instrumental in triggering responses in some greenhouse crops. ***Thermoperiodism*** is the term used to describe a temperature requirement that produces a plant response. Poinsettias and holiday cactus are examples of thermoperiodic plants. A period of cool temperatures along with short days causes them to initiate flowering.

For some floriculture crops, a period of cold temperature is required for flowering. This physiological process is known as ***vernalization***. Bulb crops, including tulips, narcissus, and Easter lilies, must undergo vernalization. They are exposed to freezing or near freezing temperatures for a number of weeks. Once the cold requirement is satisfied, the plants are moved into the greenhouse for forcing.

DIF

Growers use temperature to control the height of plants. They do this by managing the difference between daytime and nighttime temperatures. The mathematical difference between the day temperature and the night temperature is called DIF.

DIF can be positive, negative, or zero. A positive DIF results when the day temperature is higher than the night temperature. Positive DIF causes a plant to lengthen its stems. A negative DIF occurs when the day temperature is cooler than the night temperature. Plants grown under negative DIF conditions have limited stem elongation. Zero DIF is a result of identical day and night temperatures.

Table 9–1. Effect of DIF on Stem Elongation of Greenhouse Crops

Strong or Medium Effect	Little or No Effect
Begonia	Foliage plants
Carnation	Kalanchoe
Cyclamen	Rose
Chrysanthemum	Seed geranium
Fuchsia	
Impatiens	
Lilies	
Pansy	
Petunia	
Poinsettia	
Salvia	
Zonal geranium	

EXAMPLES:

Positive DIF—A day temperature of 78°F and a night temperature of 68°F equals a positive DIF of +10°F (78 – 68 = 10). Stems elongate.

Negative DIF—A day temperature of 65°F and a night temperature of 72°F equals a negative DIF of –7°F (65 – 72 = –7). Stem elongation is limited.

Zero DIF—A day temperature of 70°F and a night temperature of 70°F equals a zero DIF or 0°F (70 – 70 = 0).

9–19. These impatiens plugs have been kept compact with the application of DIF. (Courtesy, Ball Horticultural Co.)

Advantages and Disadvantages of DIF

The main advantage to DIF is stem elongation can be increased or decreased on a daily basis. Elongation is slowed with a negative DIF and speeded with a positive DIF. DIF can be used in place of growth retardants. Costs for chemical growth retardants and the labor involved in applying the retardants are reduced. Also, the potential of harming the environment with the use of chemicals is lessened.

9–20. It is difficult in warm areas, such as Florida, to apply DIF.

9–21. Fans help to increase the exchange of gases.

One disadvantage of DIF is its application is not always possible in warm regions or during the warm summer months across the country. Warm, sunny days make it more difficult for growers to reduce day temperatures within the greenhouse. Another disadvantage is DIF can result in higher heating and cooling costs to achieve targeted temperatures.

AIR

Air has carbon dioxide and oxygen that are critical for photosynthesis and respiration. Exchange of oxygen and carbon dioxide through stomata in the leaves keeps photosynthesis operating at peak efficiency. The ability of air to move in and out of the soil is important in providing oxygen for healthy root growth. Root cells must have oxygen to undergo the vital life process of respiration. Air quality is an ingredient to producing healthy plants in the greenhouse.

Carbon Dioxide Levels

When greenhouse vents are closed, CO_2 levels can drop significantly. Carbon dioxide becomes a limiting factor and photosynthesis is slowed. Ventilation of the greenhouse helps to replenish CO_2 in the greenhouse atmosphere. Supplemental CO_2 is sometimes added to the greenhouse atmosphere when vents are not open. This is accomplished with special burners for the purpose of increasing the rate of photosynthesis. Supplemental CO_2 is most effective when used to give young plants a boost. The result is higher quality crops and a shorter production time. Carbon dioxide is usually added during the hours between 9:00 a.m. and 3:00 p.m.

9–22. The sides of this structure can be lowered to increase ventilation.

Humidity, which is water vapor in the air, affects plant growth. The growth rate of plants increases under conditions of high humidity. Lush, tropical forests exist, in part, because of frequent rain and high humidity. High humidity reduces water stress of a plant so photosynthesis can function smoothly. If the humidity is low, the dryness of the air can put stress on the plant. This is especially true if soil moisture is inadequate and wilting occurs. One drawback of excessive humidity in the greenhouse is the increase of leaf and flower diseases.

Air pollution can be damaging to plants. Dust in the air can reduce light intensity, slowing photosynthesis. Chemical pollutants, such as sulfur dioxide, can actually kill plant cells or the entire plant. Greenhouse plants grown in urban settings sometimes suffer from air pollution.

9–23. Heater gases must be vented to the outside.

Greenhouse heaters must be checked regularly for proper burning of fuel. Heaters need to burn cleanly. It is extremely important to vent gases from heaters to outside the greenhouse. Gases from burning fuel are harmful to plant growth causing foliar damage. Faulty heaters or inadequate venting can create risks to workers as well.

WATER

Watering is the most important cultural practice in the greenhouse. Life processes of the crops depend on water. Photosynthesis and respiration require water. Roots are able to absorb minerals only if the minerals are dissolved in water. Water is the carrier of materials through the xylem and phloem. Water also makes up a large percentage of the plant cells, tissues, and organs.

A lack of water places plants under stress. Plants transpire water through their stomata. If the plant roots cannot supply water quickly enough to replace water lost from transpiration, the plant responds by closing its stoma. This helps the plant to conserve water. However, when closed, the stoma do not permit an exchange of carbon dioxide and oxygen. Without carbon dioxide, photosynthesis, or the manufacture of sugars, cannot occur. Severe water loss causes wilting. **Wilting** is a drooping condition and a lack of firmness to the plant tissues caused by inadequate water. Wilting is a result of a loss of **turgidity**, or water pressure, in the plant cells. Wilting is very serious with chrysanthemums. Chrysanthemums allowed to wilt severely never fulfill their full flowering potential.

9–24. The plugs in this photograph will die unless they are watered.

Water must be given to plants when needed. Timing is critical. Frequent applications of water can keep a growing medium too wet. The result may be damaged plant roots caused by the lack of good air exchange. Allowing soils to become too dry between waterings may also cause root death and lower the quality of the crop. In the greenhouse, it is not unusual for plants to need water more than once during a warm, sunny day. Therefore, the person in charge of watering plants must have knowledge of the crop being grown and good judgment about when to water. A good rule is to water plants thoroughly when they need water and wait until they need water before watering again.

Water Quality

The quality of the available water is carefully researched and tested in selecting a site for a greenhouse. Water with a favorable pH range of 5.8 to 6.2 is best for most plant growth. If the water supply has a pH outside that range, the pH can be adjusted. A second consideration in determining water quality involves soluble salts. Water with high levels of soluble salts is detrimental to plant growth. Soluble salts damage roots. Removal of soluble salts from the water supply is an expensive operation.

Irrigation Methods

Furnishing plants with water by flooding or sprinkling is known as ***irrigation***. There are a number of methods used to irrigate or water plants. With new techniques involving automation, labor has been reduced. Thousands of plants can be watered at one time. Several different watering methods are used.

Hand watering is the most reliable method of watering. However, it is time consuming and great responsibility is placed on the person performing the task. A wand is usually placed on the end of a hose with a nozzle to break the flow of water.

Watering Tips

1. Water when the plant needs water and wait until it needs water before watering again.

2. Avoid wetting the leaves and flowers. Moisture can lead to problems with fungal diseases.

3. Never allow the breaker nozzles to touch the floor. Disease organisms on the floor can be picked up and spread to the plant during watering.

4. Water thoroughly to wet all of the growing medium and to leach soluble salts from the growing medium. This sometimes means watering a container more than once.

9–25. This person is using acid to lower the pH of the water before irrigating the plants.

9–26. Although labor intensive, hand watering is the most reliable method of watering.

9–27. A spaghetti tubing system waters the gloxinias shown.

Spaghetti tubing involves small tubes connected to a main line. The end of each tube is placed in an individual pot. When operating, water dribbles through the tubes, watering all the pots on a bench simultaneously. An advantage is foliage and flowers do not get wet.

Drip irrigation is similar to spaghetti tubing except the tube provides a slow steady drip with little run-off. Less water and fertilizers are used than with spaghetti tubes. Drip irrigation is widely used with hanging baskets.

With the ***ebb and flood method***, water is pumped into the system at regular intervals, filling the bench. The pots sit in the water. The medium in the pots slowly absorbs water. After a certain period, the water drains from the bench. Ebb and flood benches conserve water, provide accurate nutrient levels, and allow for optimum spacing of plants.

Capillary mats consist of porous mats laid on a bench and wetted. The medium in the pots absorb the water by capillary action. Capillary mats are frequently used in Afri-

9–28. These tanks hold nutrient solution for an ebb and flood system.

9–29. This person programs a computer for an ebb and flood system.

9–30. The leaves of these poinsettias stay dry on an ebb and flood bench.

9–31. Capillary mats are often used to water African violets and plants in small pots.

can violet production and with other plants grown in small pots. The plant foliage stays dry and there is little runoff.

Saucers are used with overhead irrigation systems in a manner similar to capillary mats. As water is applied, it is collected in the saucer and drawn up into the growing medium by capillary action. Saucers reduce water and fertilizer waste.

Computer controlled ***irrigation booms*** sweep across greenhouse benches delivering overhead water. Large areas of bench space are watered quickly. Irrigation booms are very useful for bedding plant production. Some hand watering is needed for touch up work. Also, the foliage of the plants does get wet.

Overhead sprinklers are used primarily with bedding plants. Sprinkler nozzles in a fixed position above the crops deliver overhead water. While they water a large area in a short period, application is often uneven. Also, water and fertilizer are wasted and foliage gets wet.

9–32. Plastic saucers are used with these chrysanthemums to collect water.

9–33. An irrigation boom waters these poinsettias.

9–34. Overhead sprinklers are used primarily with bedding plants.

Intermittent Mist Systems for Plant Propagation

An ***intermittent mist system*** delivers water in tiny droplets for the purpose of keeping plant material moist. Intermittent mist systems are used for plant propagation. When taken, cuttings are without roots to absorb water. The mist system can relieve water stress on the cuttings until the cuttings develop roots. Mist systems permit higher light intensity. The higher light intensity increases sugar production and hence, speeds the production of roots.

Intermittent mist systems can operate continuously or be set to mist the plants at regular intervals. Since there is less water stress at night, mist systems are set to operate only during the day. A solenoid valve controls the flow of water through the system. The solenoid is activated by one of three control mechanisms.

9–35. Intermittent mist systems are used in plant propagation.

A time clock can be set to determine the time of day the mist system will operate. It also is set for the frequency and duration of the mists. A typical frequency of operation during the day might be 30 seconds every ten minutes.

A weighted leaf system consists of a fine metal mesh leaf that is balanced with a switch. When the leaf is dry, it triggers the switch to turn on the mist. The leaf system better reflects the conditions in the greenhouse than a clock timer does. It is more active on warm, sunny days, than cool, cloudy days.

A third controlling device involves a computer. Computer devices program the frequency and duration of the mists. They can also take into account the environmental conditions in the greenhouse.

REVIEWING

MAIN IDEAS

Growers manage light, temperature, air, and water in the greenhouse to promote healthy growth. These production practices are referred to as cultural practices.

Light plays a role in many aspects of plant growth and development. The effect of light is a result of the color of the light, the duration of the light, and the intensity of the light. Blue and red wavelengths produce a higher level of photosynthetic activity and chlorophyll synthesis than other colors. A plant's response to light duration is called photoperiodism. Plants are classified based on their response to light duration. They may be short-day, long-day, or day-neutral plants. Growth of the major floriculture crops is most active when they receive full sunlight. Growers can reduce light levels in the greenhouse when needed. They use artificial lighting to alter light duration and to increase light intensity.

Temperature is an important factor in biochemical reactions, including photosynthesis and respiration. In general, life processes are more active under warm temperatures. Thermoperiodism is a term used to describe a temperature requirement that produces a plant response. Some floriculture crops require a period of cold temperature or vernalization to flower. Temperature management is used to control the growth of plants. This is achieved by managing the difference between daytime and nighttime temperatures. The mathematical difference between the day temperature and the night temperature is called DIF. Stems elongate under positive DIF conditions and stay shorter under negative DIF conditions.

Air quality is an ingredient to producing healthy plants in the greenhouse. Carbon dioxide levels influence the rate of photosynthesis. A good air exchange between the greenhouse atmosphere and the outside atmosphere helps to keep CO_2 levels high. Sometimes, CO_2 burners are used to provide additional CO_2 in the greenhouse. Air pollution can be damaging to plants. Greenhouse heaters must be checked regularly for proper burning of fuel. It is also important to vent gases from heaters to outside the greenhouse.

Water is something that must be given to plants when it is needed. A good rule is to water plants thoroughly when they need water and wait until they need water before watering again. Automation has made watering a large number of plants in a short period possible. There are a number of methods used to water plants including hand watering, spaghetti tubing, drip irrigation, ebb and flood, capillary mats, irrigation booms, and overhead sprinklers. Intermittent mist systems are used for plant propagation.

QUESTIONS

Answer the following questions using complete sentences and correct spelling.

1. What is meant by cultural practices?
2. How do the colors of light influence plant growth and development?
3. How does light duration impact plant growth?
4. How does light intensity influence plant growth?
5. What are some applications of artificial lighting systems?
6. How do thermoperiodism and vernalization impact floriculture crop production?
7. Why is carbon dioxide sometimes added to the greenhouse environment?
8. Why is watering considered to be the most important cultural practice in the greenhouse?
9. What is a good rule for watering?
10. What are the major methods of irrigating greenhouse crops?

EVALUATING

Match the term with the correct definition. Write the letter of the term on the line provided.

a. cultural practices
b. DIF
c. ebb and flood
d. high-intensity discharge (HID) lighting systems
e. high pressure sodium lamps
f. photoperiodism
g. short-day plants
h. spaghetti tubing
i. vernalization
j. wilting

______1. The control or management of light, temperature, air, and water to promote healthy growth.

______2. Plants that will begin to flower naturally in the fall when the day lengths get shorter.

______3. A drooping condition and a lack of firmness to the plant tissues caused by inadequate water.

______4. An HID supplemental lighting system that emits a high level of red, yellow, and orange wavelengths.

______5. The mathematical difference between the day temperature and the night temperature.

______6. The physiological process of some floriculture crops whereby a period of cold temperature is required for flowering.

______7. A watering method in which water is pumped into the system at regular intervals, filling a watertight bench.

______8. A plant's response to light duration.

______9. Artificial lighting systems that have given growers more efficient and effective light sources.

______10. This watering method involves small tubes connected to a main line through which water dribbles, watering all the pots on a bench simultaneously.

EXPLORING

1. Make an appointment to visit with a grower at a local greenhouse. Be prepared to ask questions concerning cultural practices involving light, temperature, air, and water. Keep notes on the grower's responses.

2. Conduct a school greenhouse "cultural practice inventory" with your instructor. Identify practices that could be changed to improve crop production. Work to implement approved cultural practices.

CHAPTER 10

Growing Potted Flowering Crops

OBJECTIVES

This chapter is about growing potted flowering crops. It has the following objectives:

1. Identify major potted flowering crops and their origin.
2. Describe the importance of flowering potted plants to the floriculture industry.
3. Arrange a schedule for a $6^1/_2$-inch pinched poinsettia crop.
4. Set up a growing schedule for a chrysanthemum crop.
5. Describe cultural practices in producing an African violet crop.
6. Identify basic growing requirements for kalanchoes, cyclamen, and minor potted flowering crops.

TERMS

African violet
center bud removal
chrysanthemum
cineraria
cyclamen
disbudding
florist azalea
florist hydrangea
gloxinia
hard pinch
heat delay
holiday cactus
kalanchoe
liners
Persian violet
pocketbook plant
poinsettia
primrose
roll out pinch
soft pinch

10–1. The Opryland Hotel in Nashville uses potted flowering plants for a festive look during the holidays.

YOU WOULD find it difficult to go to a mall or an office building and not find potted flowering plants used in the interior landscape. They can bring bright colors to ordinarily cheerless surroundings. This is especially true around the holidays when festive looks are desired. Flowering potted plants are closely associated with each of our major holidays.

Can you think of which flowering plants are sold around Christmas? What flowering plants are sold at Valentine's Day, Easter, and Mother's Day? Have you ever wondered where the plants come from? Why are some of them available at certain times of the year and not at other times? There are many more questions that you might think about if you really give flowering potted plants some thought.

POTTED FLOWERING PLANTS

Potted flowering plants consist of plants grown in pots for their showy flowers. The major potted flowering crops are part of everyday life in America. Most people easily identify poinsettias, chrysanthemums, and African violets. Supermarkets, florists, garden centers, and chain stores sell potted flowering plants. Offices, malls, and retail stores use them to brighten their business environments.

Potted flowering plants are sold in all areas of the country. They make up a large portion of greenhouse production sales in the United States. Their segment of the greenhouse production industry is second only to bedding plant production. Based on the 2002 USDA Floriculture Crops Summary, the total wholesale value of the potted flowering crops was $495 million.

Table 10–1. Potted Flowering Crop Sales According to the 2002 USDA Floriculture Crops Summary

Crop	Total Pots Sold in 2001
1. Poinsettia	67,375,000
2. Chrysanthemum	25,834,000
3. Spring Flowering Bulbs	17,889,000
4. African violet	15,422,000
5. Florist azalea	13,963,000
6. Orchid	12,119,000
7. Florist rose	10,441,000
8. Easter lily	9,581,000
9. All others (Cineraria, Calceolaria, Cyclamen, Kalanchoe, Primrose, Persian violet, Gloxinia, Holiday cactus, Florist hydrangea, etc.)	68,141,000

POINSETTIA (*Euphorbia pulcherrima*)

General Information

Poinsettias are native to Mexico. The first U. S. Ambassador to Mexico, Joel Poinsett (1779–1851), introduced poinsettias to the United States in 1828. Until the early 1900s, poinsettias were used chiefly as cut flowers. Poinsettias are associated with Christmas, and they are now sold as potted plants only during that time of the year. The Ecke family in California promoted the use of poinsettias. To this day, the Ecke family is recognized as the leader in poinsettia production. They supply stock plants and cuttings for growers around the country.

10–2. Poinsettias are associated with Christmas. (Courtesy, Agricultural Research Service, USDA, Scott Bauer)

The first poinsettia plants were much less attractive than today's poinsettia cultivars. They dropped their lower leaves creating a leggy appearance, had small bracts, and failed to hold their color. New production techniques and hybridizing have greatly improved the appearance of poinsettias. Plant breeding has led to new cultivars of poinsettias that hold their color longer, have larger bracts, and are more compact than earlier varieties. Success in propagation has increased dramatically with mist systems. Also, growth regulators have been affective in keeping plants more compact.

Poinsettia cultivars are available in a variety of colors including the traditional deep red, pink, white, speckled, and even yellow. They are grown as single stem or branched plants. They can be trained to grow as small trees or used in hanging baskets. Pot sizes used range from 4-inch to 12-inch azalea pots.

10–3. Poinsettia cultivars are available in a variety of colors.

Table 10–2. Growing Guidelines for Poinsettia Varieties (Information provided by Paul Ecke Ranch)

Cultivar	Colors	Response Weeks	Height	E.C.	Durability	Shelf Life
EARLY SEASON RED/FAMILIES						
Pepride Family	R, W, P, M	7.5-8	Short	1.5–2	Best	Best
Freedom Family	R, Brt.R, W, Brt. P, Rose, Marble JB, Coral, Salmon	8	Medium	1.5–2	Best	Excellent
Freedom Fireworks	R	8	Medium	1.5–2	Best	Excellent
Punch Family	R, W, Lt. Pink, Rose, Cranberry	8	Medium	1.5–2	Best	Excellent
Jester	R, JB	8	Short	1.5–2	Excellent	Excellent
EARLY SEASON NOVELTY						
Jingle Bells 4.0	2-tone R/P	8	Medium	2–2.5	Very Good	Very Good
Strawberries 'N Cream	2-tone P/W	8	Short	1.5–2	Best	Best
MID SEASON RED/FAMILIES						
Prestige	R	8.5	Medium	1.5–2	Best	Best
Peterstar Family	R, W, P, M, Silverbells, Orange	8.5	Medium	2	Very Good	Very Good
Lilo	R	8.5	Medium	1.5–2	Very Good	Best
Celebrate 2	R	8.5	Medium	2	Very Good	Very Good
Supjibi Family	R, W, Burgundy	8.5-9	Medium	2–2.5	Very Good	Very Good
Max Red	R	9	Short-Medium	1.5–2	Best	Best
Red Velvet	R	9	Medium	2	Excellent	Excellent
Bright Red Sails	R	9	Medium	1.5–2	Very Good	Excellent
Red Satin	R	9	Medium	2–2.5	Excellent	Excellent
Gutbier V-17 Angelika Family	R, W, P, Marble	9	Tall	2	Very Good	Very Good

(Continued)

Table 10–2 (Continued)

Cultivar	Colors	Response Weeks	Height	E.C.	Durability	Shelf Life
MID SEASON NOVELTY						
Marblestar	Marble	8.5	Medium	2	Excellent	Very Good
Maren	P	8.5	Medium	2	Excellent	Very Good
Snowcap	W	8.5	Medium-Tall	2	Excellent	Excellent
Plum Pudding	Purple	9	Tall	2–2.5	Very Good	Excellent
Holly Point	R	9	Short	1.5–2	Very Good	Very Good
Amazone Peppermint	2-tone P/P	9	Medium	2	Very Good	Very Good
Heirloom Family	R, P, Peach	9	Short	2–2.5	Very Good	Very Good
LATE SEASON RED/FAMILIES						
Winter Rose Family	R, W, P, Deep Pink, Marble, JB	9.5	Small-Medium	1.5–2	Best	Best
Success Family	R, Lt. Pink, Coral	9.5	Medium	2–2.5	Best	Excellent
Gutbier V-14 Family	R, W, P	9.5	Medium	2–2.5	Very Good	Very Good
Red Splendor	R	10	Medium	1.5–2	Best	Very Good
LATE SEASON NOVELTY						
Monet, Monet Twilight	2-tone R/P	9.5	Medium	2–2.5	Best	Excellent
Candy Cane	2-tone P/W	9.5	Medium	2–2.5	Very Good	Very Good
Jingle Bells, Jingle Bells 3	2-tone R/P	9.5	Medium	2–2.5	Very Good	Very Good

Photoperiodic Response

It is important to understand that poinsettias are photoperiodic. As short-day plants, they initiate flower buds as days get shorter. Flower bud initiation begins when nights reach 11 hours and 50 minutes of darkness. That naturally occurs during mid- to late-September in the central portion of the country. September 25 is considered the key date for flower bud initiation. The day length reaches the critical 11 hours and 50 minutes earlier in the north. As one progresses toward the south, the 11 hour 50 minute day is later in the calendar year. Poinsettias are also thermoperiodic. Lowering night temperatures in the greenhouse to 62 to 64°F contributes to flower bud initiation.

Poinsettia varieties respond differently in terms of the number of weeks of flower bud initiation until flowering. Those that have a short response may require only 8 weeks, while those with a longer response require 10 weeks. Ten week varieties are less sensitive to light at night than 8 week varieties. Growers consider the photoperiodic response in selecting the poinsettia cultivars.

Propagation

Poinsettias are propagated asexually. Growers can buy stock plants and take their own stem cuttings. They can also purchase unrooted cuttings, callused cuttings, or rooted cuttings. Unrooted cuttings have a length of 2 to 4 inches and they take about 4 weeks from sticking to rooting. Callused cuttings are $3^{1}/_{4}$ to 4 inches in length. Root initials have developed prior to shipping, and rooting takes 1 to 2 weeks less than unrooted cuttings. Rooted cuttings have a maximum height of 6 inches. Rooted cuttings become established quickly once potted.

Most unrooted cuttings are stuck near the end of July as part of the overall crop schedule. A suggested procedure follows:

- Cuttings are taken 2 to 4 inches in length.

10–4. Some growers buy stock plants in February, pinch them to induce branching, and then take cuttings.

- Dip or dust the end of the cuttings in a medium-strength rooting hormone and stick the cutting in a rooting cube or directly in the finish pot.
- Place under intermittent mist for 24 hours the first day. Adjust to 15 to 20 seconds every 3 to 5 minutes during the first week and 10 seconds every hour at night. Mist permits higher light intensity, which helps the plants produce more sugars for root development.
- Reduce the period of mist to 10 seconds every 10 minutes as the cuttings form root initials.
- Provide a temperature range of 70°F to 75°F during rooting.

Potting

Poinsettias require a well-drained and well-aerated growing medium with a pH between 5.0 and 6.5. The recommended pot for most production is the azalea pot. Soilless medium settles so pots should be filled to the top. It is important for all of the pots to have the same amount of medium. This helps with watering practices and in uniform development of the entire crop. Plant the cuttings in the center of the pot and at the same depth. Shallow placement promotes healthy root growth. This is because the top few inches of medium has 20 to 30 percent more air than the lower medium. The root cubes can be exposed because there is little concern about wicking moisture from the medium. After planting, drench the pots with fungicide to control root and stem rot.

10–5. This worker pinches plants in mid-September.

Table 10–3. Cuttings Per Pot and the Number of Days Needed for Them to Become Established (Information provided by Paul Ecke Ranch)

Number of days required to establish rooted cuttings into container prior to pinch.		
Container Size	Number Cuttings per Container	Number of Days
3"	1	5
4–4½"	1	8
5½–6"	1–2	10–14
6½"	1–3	14
7"	1–5	14–17
8"	1–7	17–20
10"	1–10	20–25

Note: Exceptions to this guideline include: The production of tree forms that require extended growing prior to the initial pinch; the production of specific cultivars like Eckespoint Lilo and its color sports that require pinching within 10–14 days of transplant.

Branched Versus Single Stem

Poinsettias are grown as branched or single stem plants. Branched poinsettias are the most commonly grown form today. To produce a branched plant, poinsettias are pinched. By pinching off the growing tip, lateral buds in the axils of the remaining leaves develop. It is advised to leave 4 to 6 leaves. A flower will develop at the end of each branch that develops. Hence, a single cutting can produce multiple flowers.

An alternative to branched plants is non-pinched plants in a pot. A lone flower is produced at the terminal end of the plant. For the best display, a number of cuttings must be planted in each pot. The production time for non-pinched plants is shorter when compared to pinched plants. Therefore, plants can be started later. Symmetry and uniformity are easier to achieve. Also, there tends to be less breakage during shipping and handling.

Stages of Development

Poinsettia production can be broken down into four stages of development. They are the vegetative stage, flower bud initiation, flower bud development, and flowering. Lighting and temperatures are adjusted according to the stage. The fertilization schedule also changes with the production stage.

The first thing a grower should do is to see that the poinsettias cuttings are well established in the pot. They should develop a root system that has reached the outsides of the pot. Leaves should also indicate healthy growth. Strong early growth contributes to stored sugars needed to produce branching of pinched plants. Plants grown in 4 inch pots require only eight days from potting to pinching. Those that are to be larger plants grown in larger pots need 10 to 25 days. Refer to Table 10–3.

Table 10–4. Days of Vegetative Growth Prior to Flower Bud Initiation (Information provided by Paul Ecke Ranch)

Number of days vegetative growth suggested between pinching and flower initiation as determined by product form and geographic location.

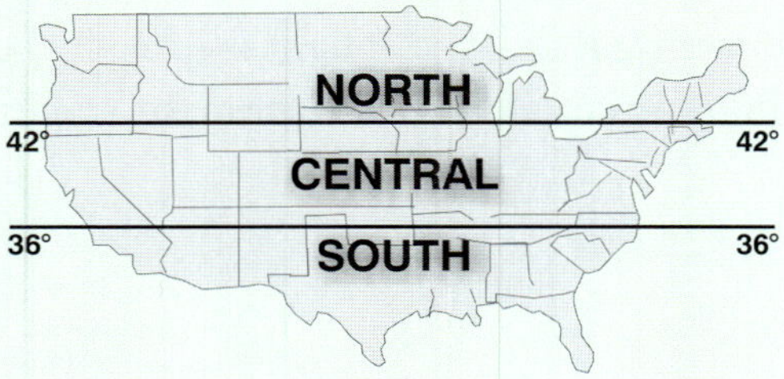

Product Form	Geographic Location		
	South	Central	North
3"	0	3	5
4–4½" 1 Plant Pinched	5	7	9
5½–6" 1 Plant Pinched	9	12	15
6½" 1 Plant Pinched	12	16	22
7" 2 Plant Pinched	18	22	27
8" 2–3 Plant Pinched	21	26	31
10" 3–4 Plant Pinched	25	30	35

Table 10–5. Additional Days of Vegetative Growth Based on Poinsettia Cultivar's Growth Habit (Information provided by Paul Ecke Ranch)

Number of additional days vegetative growth suggested between pinching and flower initiation as determined by growth habit characteristics.

Product Form	Growth Habit		
	Short	Medium	Tall
3"	5	0	0
4–4½" 1 Plant Pinched	5	5	0
5½–6" 1 Plant Pinched	10	5	0
6½" 1 Plant Pinched	10	5	0
7" 2 Plant Pinched	10	5	0
8" 2–3 Plant Pinched	15	10	0
10" 3–4 Plant Pinched	15	10	5

Poinsettias will grow vegetatively as long as the days are long. During the vegetative development of the plant, it is important to maintain long day conditions. To accomplish this it may be necessary to interrupt the night with artificial lighting. Lighting between 10 p.m. and 2 a.m. each night between August 1 and September 25 keeps the plants in a vegetative state of growth.

Short day conditions send a signal to the plant to begin initiating flower buds. This is important since no bract coloration occurs without flower bud initiation. Flower bud initiation occurs over a period of 8 to 10 days. Growth of the plant stops once the plants initiate flower buds. After that point, cells just enlarge. Giving plants more time between pinching and flower bud initiation results in a larger finished plant.

In the third stage, flowers develop. Flower bud development takes place through much of October and the first couple weeks of November. Flowering is delayed if light is introduced to the plants, even for short periods, during flower bud initiation and development stages. After mid November, the plants are finished for sale.

Growing Schedule

A growing schedule for 6 to $6^{1}/_{2}$-inch multi-flowered poinsettia crop in soilless mix in the central geographic area of the United States follows.

Late August—Vegetative Stage

Propagate cuttings between July 25 and August 5. Pot poinsettias in a finishing pot between August 20 and September 1. Drench with fungicides to reduce problems from soil borne diseases. Promote vegetative growth by lighting plants from 10 p.m. to 2 a.m. Fertilize

10–6. These plants were just pinched leaving six leaves.

10–7. With the removal of the apical meristem, side shoots or breaks develop.

with 300 ppm nitrogen and potassium. Provide 68 to 70°F night temperatures and 70 to 80°F day temperatures.

September 10—Vegetative Stage

Drench with Cycocel, a growth regulator, if height is a concern. Pinch the plant leaving four to six leaves between September 5 and September 15. Maintain high humidity to encourage breaks. Note, that the longer the period between pinching and flower bud initiation, the larger the plant. Raise fertilizer rate to 350 to 400 ppm nitrogen and potassium. Maintain night temperatures of 68 to 70°F.

10–8. This photograph shows automatic short day curtains being closed over a crop of poinsettias to ensure short days and long nights.

September 20 to 25—Flower Bud Initiation Stage

Turn lights off and give plants short days. If nearby lights interfere with the natural darkness, cover the crop with short day curtains from 5:00 p.m. until 8:00 a.m. Drop night temperatures to 62 to 64°F and day temperatures to 70 to 72°F. Fertilize at 300 to 350 ppm nitrogen and potassium. Initiation of flower buds occurs over a period of 8 to 10 days. Cell division stops after flower buds initiate. From then on, the cells enlarge. Space plants 15 inches by 15 inches to avoid stretching.

10–9. Throughout the fall, this grower spaces plants so they are not shading one another.

10–10. A finished plant has a flower at the end of each stem that developed after pinching.

10–11. Whitefly is the most serious pest on poinsettias. (Courtesy, Agricultural Research Service, USDA, Scott Bauer)

October 10—Flower Bud Development Stage

Flower buds begin to develop. Stop black cloth treatment. Adjust temperatures to 64 to 66°F at night and 70 to 75°F during the day. Fertilize at a rate of 300 ppm nitrogen and potassium. Drench with a fungicide, such as Banol 66.5%, Banrot 40W, Chipco aliette 80 WDG, Subdue II WSP, Terrazole 35 WP, or Truban 25EC or 30 WP to control root rot.

November 15—Flowering Stage

Growers can increase the life of poinsettias by finishing them in preparation for shipping to the customer. Begin to finish plants around November 15. Drop temperature to 58 to 62°F nights to deepen bract color. Reduce fertilizer rate to 200 to 300 ppm nitrogen and potassium. Also, maintain good soil moisture management.

Poinsettias are very sensitive to cold temperatures. Placing plants in sleeves offers some protection from cold and physical damage during shipping. Plants can be stored in plant sleeves for a short while at 60 to 65°F.

Pests and Disease Problems

The most serious insect pest on poinsettias is the whitefly. Mites, mealy bugs, thrips, and fungus gnats can also create problems. A clean, weed-free greenhouse and a pest control program reduce insect problems. Botrytis is a problem in the propagation area. Poinsettias are also very susceptible to three major root rot diseases, *Pythium*, *Rhizoctonia*, and *Thielaviopsis*. Well-timed fungicide applications reduce losses caused by fungi.

CHRYSANTHEMUM (*Dendranthema grandiflora*)

General Information

Chrysanthemums (mums) have their origin in China, Japan, and Europe. There is evidence that chrysanthemums were used in Chinese gardens around 550 BC. Mums were brought to the United States and used only in the fall until the 1940s.

10–12. A chrysanthemum in flower.

Then, it was found that chrysanthemums are photoperiodic and thermoperiodic. Because flower buds set naturally from middle August to late September, chrysanthemums are considered to be short-day plants. They are lighted for vegetative growth and shaded between March and September to induce flowering. By controlling light duration and temperatures, potted mum crops can be produced throughout the year. Easter and Mother's Day are two holidays that see high sales of potted chrysanthemums.

10–13. The label for this crop indicates when it was planted, when short-days are to begin, and when it will flower.

Most chrysanthemum varieties initiate flower buds with twelve or fewer hours of daylight. A few will flower with only $11^1/_2$ hours of uninterrupted darkness.

Chrysanthemum varieties are very diverse. Mums come in many colors and the flower forms differ. There are also physiological differences that influence production schedules. Mums are classified based on these differences:

- **Response group**—One grouping is based on the number of weeks it takes them to flower from the time they begin receiving short days. For instance, a nine-week variety takes nine weeks to flower once it is given short-day treatments. The response groups of mums range from 6 weeks to 15 weeks. Garden mums include six-, seven-, and eight-week varieties. The majority of potted chrysanthemums have a 9- or 10-week response.

- **Plant height**—A second grouping involves the height of the plants. Chrysanthemum varieties are identified as being tall, medium, or short.

  Tall varieties grow taller than 15 inches in height. They may receive short-day treatment prior to pinching.

  Medium varieties are around 15 inches tall and receive short-day treatment at the time of pinching.

  Short varieties grow less than 15 inches in height and need an additional 7 to 14 days growing time after pinching before given short-day treatment.

- **Flower forms**—Mums are also grouped by their flower forms, which include standard, spray, button, spoon, decorative, daisy, spider, pompon, and anemone.

10–14. Chrysanthemums have different flower forms. (Courtesy, Yoder Brothers, Inc.)

Propagation

As the popularity of chrysanthemums grew in the 1940s and 1950s, it was discovered that many of the plants were infected with viruses that produced deformed growth. The Yoder Brothers Company began a crusade to clean infected plants from the industry. They began producing cuttings free of virus and other systemic diseases. Today, a few specialized companies provide over 95 percent of all the chrysanthemum cuttings. The cuttings are sold as rooted cuttings or as unrooted cuttings.

10–15. Chrysanthemums can be purchased as unrooted cuttings. These are dusted with a rooting hormone.

10–16. These cuttings were graded before planting to get uniform growth.

Cuttings are taken $2^1/_2$ to 3 inches in length with three leaves. They are rooted with bottom heat of 70 to 75°F and lit to maintain vegetative growth. Intermittent mist is provided until roots form. Chrysanthemums are basal rooters. Fast-rooting varieties root in about 18 days.

Growing Procedures

The cuttings are graded or sorted before planting. This is done according to the number of roots, diameter of the stem, and the length of the stem. Failure to grade cuttings results in pots with lopsided growth. Some cuttings in the pot may grow vigorously while others grow more slowly.

Chrysanthemums are grown as single stem or pinched plants. They prefer a growing medium with a pH of 6.0 to 6.5. Azalea pots are the pots of choice for most growers. Typically, one cutting is used in a $4^1/_2$-inch pot. Three, four, or five cuttings are used in a 6- or $6^1/_2$-inch pot, six to seven cuttings in a 7-inch pot, and 9 to 12 cuttings in an 8-inch pot. Plant cuttings shallow and at a 45-degree angle with the tops extending over the lip of the pot. Planting cuttings too deep is a major problem. Also, the roots should not be pressed down when potting.

There are four main periods of growth and development in producing a potted chrysanthemum crop. The first phase or vegetative phase involves the promotion of root and leaf growth. The vegetative phase lasts three to five weeks depending on the variety and the time of the year. Flower bud initiation, flower bud development, and the finishing stages follow. Each phase requires a different fertilizer rate and temperature.

Vegetative Phase

The vegetative phase is the most critical period of growth. During this time, plants must build up reserves of sugar and produce a strong root system. Begin feeding newly planted rooted cuttings immediately. Use a starter fertilizer with 250 to 350 ppm nitrogen at the first watering. Starter fertilizers are typically high in phosphorous to encourage root growth. Thereafter, use 200 to 300 ppm nitrogen and potassium with each watering. Water when the growing medium approaches drying. Provide 65 to 70°F night temperatures, 75 to 80°F day temperatures, and high humidity. HID lighting in the darker months and carbon dioxide burners help the cuttings get off to a fast start.

Long days must be provided to keep chrysanthemums in a vegetative stage of growth. Interrupt the night darkness from December through February by lighting the plants from 10:00 p.m. to 2:00 a.m. From September through November and March through May, light for three hours. In the summer months, light for one to two hours to be sure plants stay vegetative.

PINCHING

Grow for 7 to 14 days or until roots appear at the bottom of the pot before pinching the plants. Pinching is the removal of the growing point of the stem. Pinching encourages breaks or branching of the plants. Three to four strong breaks from a pinch are desired. Plants with good lower leaf development produce more breaks than weaker plants. The results of pinching include better-shaped plants that appear fuller and produce more flowers. There are three methods of pinching:

10–17. There are three methods of pinching.

10–18. This grower is taking a soft pinch.

The ***roll out pinch*** involves the removal of just the meristematic tip of the stem. Care needs to be taken to remove the whole tip or growth from the tip resumes. The roll out pinch is recommended in winter when growth is sluggish.

A ***soft pinch*** involves the removal of 1/2 inch to 3/4 inch of the stem. Five to six leaves are left on the cutting. This method is used throughout most of the year.

The ***hard pinch*** removes everything but three to four leaves. The hard pinch is used in the summer with tall cuttings.

Flower Bud Initiation

Turn off night lighting to provide short-day treatment for the initiation of flower buds. Cover the crop with black cloth if necessary to extend the period of light duration. The desirable temperature range for flower bud initiation is 62 to 65°F nights. Also, lower the humidity levels in the greenhouse. Reduce fertilizer rates to 200 to 250 ppm.

In the summer months, temperatures can build up under black cloth resulting in heat delay. ***Heat delay*** is caused by day temperatures over 90°F or night temperatures over 80°F. The result is a delay of flowering by several days to weeks.

Use of a growth retardant or DIF effectively controls the height of tall varieties. Typically, a foliar application of B-Nine is applied 7 to 18 days after pinching or when about 2 inches of new growth has occurred. Short varieties and nine-week varieties do not need growth retardant applications. A slightly negative DIF of –1 to –4 has proven effective.

Flower Bud Development

At night, drop temperatures to 60 to 62°F. Continue to fertilize at a rate of 200 to 250 ppm.

As the flower buds develop, they may be removed to improve the overall quality of the plants. This practice is known as ***disbudding***. One method of disbudding is ***center bud removal***, which involves the pinching off of the most terminal flower bud. This allows the lateral flower buds to develop. Center bud removal is performed with decorative-type

10–19. Disbudding involves the removal of flower buds.

10–20. With the lateral flower buds removed, all the energy goes into the terminal flower buds.

mums. Another disbudding practice involves the removal of all the lateral buds. The plant responds by sending all its energy to the remaining bud. The result is larger showier flowers.

Finishing

Drop temperatures to 55 to 58°F at night to increase color intensity and quality. Growth of the plant at this point has stopped. Stop fertilization and allow the growing medium to dry a bit more than earlier in the crop before watering.

A quality final product will be symmetrical and have all the flowers opening at the same time. A florist-quality potted mum in a 6$^1/_2$-inch pot will follow a 15-15-15 rule of thumb. That is, it is 15 inches tall from the lip of the pot, 15 inches in diameter, and it has 15 flowers.

10–21. This florist-quality mum nearly ready for sale is 15 inches tall from the lip of the pot, 15 inches in diameter, and it has 15 flowers.

Scheduling

Classifying chrysanthemums by response group has made timing of crops for a certain sales date easier. We can look at an example schedule for a Valentine's Day crop using a nine-week, medium-size variety. For marketing purposes, the plants need to begin flowering by February 7. Begin by counting backwards on the calendar nine weeks from February 7. This takes us to the date when plants are pinched and short-day treatment is begun. Count back another two weeks for vegetative growth prior to pinching. The total time for production of a 9-week variety from rooted cuttings is 12 weeks. So, rooted cuttings should be ordered for and planted 13 weeks before February 14.

Table 10–6. Simplified Growing Schedule

Tall varieties	Vegetative 1 week	Short days 1 week	Pinch	
Medium varieties	Vegetative 1–2 weeks	Pinch	Short days	
Short varieties	Vegetative 2 weeks	Pinch	Vegetative 1 week	Short days

Pest and Disease Problems

The major pests of chrysanthemums are aphids, whiteflies, spider mites, leaf minors, cutworms, and cabbage loopers. Disease problems include botrytis, mildews, and soil-borne disease including *Pythium*, *Rhizoctonia*, and *Phytophthora.*

10–22. A granular pesticide has been placed in the pot of this chrysanthemum.

AFRICAN VIOLET (*Saintpaulia ionantha*)

General Information

African violets are native to areas south of the equator on the African continent. They are found on rainy hillsides where temperatures are warm. They are common in many American homes because of their bright and abundant flowers. Flower colors fall in the white-pink-purple range. They are produced for Valentine's Day, Easter, and Mother's Day. African violets are not photoperiodic, but they do develop more quickly under long-day conditions.

10–23. African violets are very popular plants for the home.

Propagation

African violets are usually propagated by leaf-petiole cuttings. Select half-mature leaves. They have active hormone production and the potential for growth still exists. In fact, the leaves will grow while it is rooting. It is desired to have eight to ten plantlets develop from each leaf. Tissue culture is another technique used in propagating African violets.

Leaf petioles are stuck at an angle 1 to $1\frac{1}{2}$ inches into a light medium. Temperatures are kept at 70 to 75°F. Artificial light is provided at 1200 to 1500 foot candles. Roots form in about three weeks. Plantlets form in three to four weeks.

10–24. African violets are usually propagated by leaf-petiole cuttings.

Growing Procedures

The plantlets are carefully separated so only one plantlet is in a pot. Plant African violets in a peat-based growing medium. Care must be taken not to cover the crown of the plant. The pH should be around 6.0. They are typically transplanted to the 4-inch plastic azalea pots for finishing. African violets have a very fine root system so larger pots are avoided.

Capillary mats and ebb and flood watering systems are applied in African violet production. These systems keep the plant leaves dry and provide humidity around the developing

10–25. An ebb and flood system is used to water these African violets.

plant. Provide 100 to 150 ppm nitrogen and potassium at every watering. Apply less fertilizer when light intensity is low between October and March. Also, African violets should never be allowed to dry out.

Water overhead every four to five waterings to leach soluble salts from the growing medium. However, care must be taken not to wet African violet leaves with cold water. Cold water causes enzymes in the plant cells to coagulate. Leaf tissues are damaged and white spots are created. Never apply overhead water that is more than 5 degrees cooler than the air temperature.

African violets perform best when light levels are between 1,000 and 2,000 foot candles. Below 1,000, they produce large leaves and no flowers. Above 2,000, the leaves are bleached. In many greenhouse operations African violets are given some shade. The preferred temperature range is 65 to 75°F nights.

African violets take different amounts of time to reach marketable size depending on the time of year. In general, total production time from the leaf cutting to sale is eight to nine months. In northern states, it takes 12 to 14 weeks to produce a Valentine's Day crop and 8 to 10 weeks for a Mother's Day crop. A quality plant has a single crown with leaves that are radially symmetrical. The flowers are held above the leaves to the center of the plant.

Pest and Disease Problems

Pests include cyclamen mites, root aphids, and mealy bugs. Nematodes can be a serious problem as African violets are highly susceptible. Phytophthora root rot and mildew are disease problems.

FLORIST AZALEA (Rhododendron obtusum, Rhododendron simsii)

General Information

Two species of *Rhododendron* are grown as florist azaleas. *Rhododendron obtusum* is native to Japan and *Rhododendron simsii* is native to China. Neither plant is considered to be very hardy. Most florist azaleas are grown in Florida, Alabama, Mississippi, and California where temperatures are warm and light intensity high. Different varieties are grown for Christmas, Easter, and Mother's Day. Most growers buy prefinished plants and force them to flower.

Propagation

Florist azaleas are woody plants and as such take anywhere between two and three years from propagation to flower. Stem cuttings, $2^1/_2$ to 3 inches in length, are rooted in a peat mix. The rooting process takes about 6 to 12 weeks. The rooted cuttings are pinched four to five times to encourage branching. The resulting small-branched plants are sold as 4-, 6-, or 8-inch diameter ***liners***.

10–26. A florist azalea. (Courtesy, Ball Horticultural Co.)

Growing Procedures

Received in May, liners are planted in 6-inch azalea pots in pure peat moss. Shredded pine fibers and pine bark humus are alternative medium components. Florist azaleas require an acid medium with a pH ranging from 4.5 to 5.5. The medium should never be allowed to dry completely.

Plants are pinched in June. Chemical pinching agents are sometimes used to kill the apical meristems resulting in breaks. During the vegetative stage of production, they are given 65°F night temperatures and 75 to 85°F day temperatures.

High light intensity is desired for production. Azaleas also respond to light duration. They grow vegetatively under long-day conditions. They will initiate flower buds when given short days and 60 to 65°F nights. This process of flower bud initiation takes four to six weeks.

Florist azaleas must be cooled to satisfy flower bud dormancy. Middle to late season varieties are cooled at temperatures less than 40°F for 10 to 12 weeks. If placed in a cooler, the plants need to be lighted to prevent leaf drop. After the cooling period, they are given 50 to 55°F temperatures to force flowering.

10–27. These florist azaleas are being held at 40°F to satisfy flower bud dormancy.

Application of gibberellic acid is used to help the plants overcome their dormancy. Gibberellic acid promotes the uniform opening of flowers desired of high-quality florist azaleas. Foliar applications of gibberellic acid can be used as a substitute for the period of cold treatment with some varieties.

Pest and Disease Problems

Pest problems are minimal for the grower who has florist azaleas for a few months of forcing before sale. Major pests include leafminers, spider mites, and aphids. Diseases include *Phytophthora*, *Cylindrocladium* blight and root rot, and *Botrytis* during storage.

MINOR POTTED FLOWERING CROPS

KALANCHOE (*Kalanchoe blossfeldiana*)

General Information

Kalanchoes are succulent plants with thick, fleshy leaves. They are native to Africa and Asia. Their flowers range in color from bright yellow to orange to hot pink. Kalanchoes are short-day plants.

10–28. A kalanchoe plant.

10–29. These cuttings were taken from a stock plant.

Propagation

Maintaining disease-free stock plants is very difficult, so most growers purchase rooted or unrooted stem cuttings from specialty growers. Stem cuttings, $1\frac{1}{2}$ to $2\frac{1}{2}$ inches long, are stuck without the use of rooting hormones. Cuttings root in one to two weeks. The cuttings are lighted to promote vegetative growth.

Growing Procedures

Plant kalanchoes in a well-drained medium with a pH between 6.0 and 6.5. Provide 65 to 68°F night temperatures. Higher temperatures result in taller plants. Kalanchoes respond well to CO_2 enrichment during periods of lower light.

10–30. Spaghetti tubing is used to deliver water and fertilizer solution to these kalanchoes.

Place kalanchoes on a constant feed program of 300 to 400 ppm nitrogen and potassium at the time of planting and continue until the start of short-day treatment. Then, reduce fertilizer rates to 150 to 200 ppm nitrogen and potassium. Kalanchoes respond well to ebb and flow, spaghetti tubing, and capillary mat watering systems.

Plants in 4-inch pots require no pinching, while those in 5- or 6-inch pots should be pinched two weeks after planting. Plants in the larger pots benefit from a second soft pinching when new shoots are 2 to $2^1/_2$ inches long. B-Nine is used after the first pinch to keep plants compact.

Allow two weeks after the final pinch before the start of short-day treatments. The crop takes 8 to 10 weeks from the beginning of short days to flower.

Pests and Diseases

Major pests include aphids, cabbage loopers, and leaf rollers. Crown rot, root rot, and powdery mildew are the primary disease problems.

10–31. A cyclamen plant. (Courtesy, *Greenhouse Product News*)

CYCLAMEN (*Cyclamen persicum*)

General Information

Cyclamen are attractive plants native to the Mediterranean. Their unusual-shaped flowers come in white, pink, carmine, and red. Cyclamen are considered a cool crop. As such, they have been produced in greatest numbers for winter and spring sales.

Propagation

Sexual propagation is used with cyclamen. The first of four stages of the propagation process begins with seeds being sown in plug trays. High humidity, darkness, and 63 to 65°F temperatures are provided for three to four weeks. In stage two, the developing hypocotyl (similar in appearance to a radish) is covered with vermiculite for two weeks. The vermiculite is moistened and humidity is kept high. In the third stage, seedlings are given higher light levels and temperatures of 65 to 68°F for six to eight weeks. The final step is the transplanting of plugs into 4-inch pots.

10–32. Cyclamen are grown and sold as plugs.

10–33. Note that during potting the hypocotyl is placed above the medium.

Growing Procedures

Cyclamen prefer a peat-based medium with a pH of 5.5 to 6.3. Plants should be planted high to prevent rot of the hypocotyl, which develops into a tuber. Do not fertilize the crop until the roots reach the end of the pot. Start fertilizing with 100 to 150 ppm nitrogen. Maintain temperatures between 65 and 68°F.

Transplant the plants to 5- or 6-inch finishing pots after 16 to 18 weeks. Fertilize at a rate of 200 to 250 ppm nitrogen. Maintain temperatures at 65 to 68°F. Never let the medium dry completely or the tuber is damaged. Keep the greenhouse humidity high.

10–34. Cyclamen are grown with their leaves almost touching to maintain high humidity.

Cyclamen are sensitive to high light intensity. They prefer 4,000 to 6,000 foot candles. In northern regions, they should be grown in full sun from November through March. Thirty percent shade is recommended from April through May and September through October. Fifty percent shade is needed June through August. In southern regions, more shading is required.

Flower bud development occurs after the plants have become well established and have 15 to 40 leaves. At this time, reduce fertilizer rates to 100 ppm nitrogen and potassium. Drop night temperatures to 60 to 62°F. The lower temperatures increase the number of flowers and keep the flower stalks firm. Gibberellic acid is sometimes used to speed flowering.

Pests and Diseases

Cyclamen mites, spider mites, aphids, and thrips cause damage to flowers and foliage. Disease problems include *fusarium* wilt, *botrytis*, and *phytophthora*.

Minor potted flowering plants collectively account for a large portion of sales nationwide. Some of the more common minor crops include cineraria, calceolaria, Persian violet, gloxinia, holiday cactus, hydrangea, and primrose.

GLOXINIA (SINNINGIA SPECIOSA)

The **gloxinia**, native to Brazil, is in the same plant family as African violets. Its large trumpet-shaped flowers come in red, pink, purple, white, and two-tones. Gloxinias are propagated from their tubers or by seed. Most growers buy seedlings from specialty growers. Plant gloxinias in a peat-based medium with a pH around 6.0. Keep the medium moist. Capillary mats, ebb and flood, and spaghetti tubing watering systems reduce the chance of damaging leaves with cold water. Provide 150 to 200 ppm nitrogen and potassium. They prefer 3,000 to 4,000 foot candles so shading is usually required. Gloxinias like 65 to 70°F nights and a humid greenhouse. Seed to flower takes about six months for a spring crop.

10–35. A gloxinia ready for sale.

10–36. This photograph shows leaf damage on gloxinias caused by cold overhead water.

HOLIDAY CACTUS

Three types of holiday cacti are available: Thanksgiving (*Schlumbergera truncata*), Christmas (*Schlumbergera bridgesii*), and Easter (*Rhipsalidopsis gaertneri*). The natural photoperiodic response of these plants brings them to flower around the holiday for which they are named. **Holiday cacti** are both short-day plants and thermoperiodic. Holiday cacti are propagated by cuttings taken from stock plants. The medium must have good drainage and a pH between 5.5 and 6.0. Fertilize holiday cacti with 100 to 150

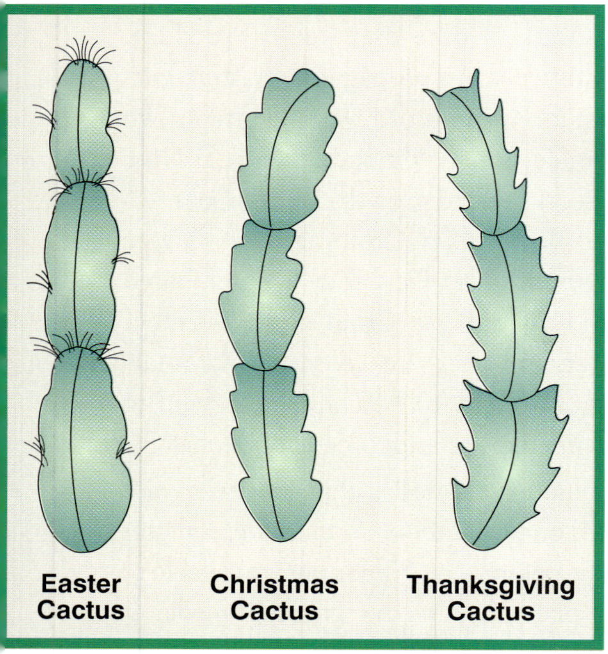

10–37. Three types of holiday cacti.

10–38. These people are taking holiday cactus cuttings.

10–39. These Thanksgiving cacti will flower naturally around Thanksgiving.

ppm nitrogen and potassium with each watering up until one month before flower buds develop. Thanksgiving and Christmas cacti initiate flower buds at 55°F nights. Easter cactus initiates flower buds when given night temperatures between 47 and 53°F.

FLORIST HYDRANGEA (HYDRANGEA MACROPHYLLA)

Florist hydrangeas are native to Japan. They are grown primarily for Easter and Mother's Day. Their long-lasting flowers come in blue, pink, white or red. Production has dropped because of the space they require and the length of time they take to produce. Hydrangeas are propagated by stem cuttings in April and May. They are grown as single stem or pinched plants. B-Nine is used to retard growth. Hydrangeas initiate flower buds from middle to late August. In October, they are defoliated with ethylene gas. From mid-November to January they are given a cold treatment of 35 to 40°F. They are moved to the greenhouse and forced at 60 to 64°F.

10–40. A florist hydrangea in flower.

Blue-pink varieties change flower color based on fertilization. Aluminum promotes a blue coloration. Aluminum is available to the plants when the growing media is acidic. Growers will provide medium with a pH of 5.5 or less for blue flowers and raise the pH above 6.5 to produce pink flowers.

10–41. Florist hydrangea produce vegetative growth until flower buds begin to develop in late August. (Courtesy, *Greenhouse Product News*)

10–42. This grower adds aluminum sulfate to his hydrangea crop to promote blue coloration.

PERSIAN VIOLET (EXACUM AFFINE)

Persian violet is from the Island of Socotra off the tip of Somalia. It has pretty violet-blue flowers that cover the whole plant. Most Persian violets are grown in 4- to $6^1/_2$-inch pots. They are propagated by seed or cuttings. Persian violets are grown for

10–43. A pH below 5.5 produces blue hydrangea flowers. (Courtesy, California Cut Flower Commission)

10–44. Persian violet.

sales ranging from Mother's Day through October. They are long-day plants that like warm temperatures. Preferred growing temperatures are 60 to 65°F nights and 72 to 77°F days. Moderate fertilization rates of 150 to 200 ppm nitrogen and potassium are effective. Since they like 4,500 to 6,000 foot candles, shading is often required. They are very susceptible to *Pythium* root rot and *Phytophthora*.

10–45. A cineraria.

COOL CROPS

Northern growers produce three cool crops that share similar production schedules, cultural requirements, and usually the same greenhouse. Production is for Valentine's Day, Easter, and Mother's Day. They are the cineraria, pocketbook plant, and the primrose.

10–46. A pocketbook plant, also known as slipper flower.

10–47. A species of primrose. (Courtesy, *Greenhouse Product News*)

10–48. These people unload primrose plugs for planting.

10–49. These primrose are grown on the floor of this poly house where temperatures are cooler.

Native to the Canary Islands, the **cineraria** (*Senecio cruentus*) has daisy-form flowers held in a cluster above the foliage. They come in blues, pinks, reds, and white. The native range of the **pocketbook plant**, also known as slipper flower (*Calceolaria herbeohybrida*), extends from Mexico to Chile. Its flowers resemble a woman's purse. The flowers are very colorful yellows, oranges, and reds. There are six species of **primrose** (*Primula* species) used in greenhouse production. All of these originate in the northern temperate regions of the world. Primroses are available in a wide range of bright, colorful, and fragrant flowers.

10–50. These cineraria are being grown in 6-inch pots. Note the level of growing medium.

Seeds of all three plants are sown in late summer. Soilless media are recommended with a pH around 6.0. Cinerarias and pocketbook plants are grown in 5- or 6-inch azalea pots, while primroses are grown in 4- to $4^1/_2$-inch azalea pots. Plants are started with 100 ppm nitrogen and potassium. Once in the finish pots, a constant feed program of 200 ppm nitrogen and potassium is recommended. During the vegetative stage, lasting four to six weeks after planting, temperatures are kept at 60 to 62°F nights, 70 to 75°F days. Shade is usually required to limit light to 4,000 to 5,000 foot candles.

Flower bud initiation is promoted after the foliage has grown to extend 2 to 3 inches over the lip of the pots, usually around mid-November. Temperatures are dropped to 45 to 55°F nights. Day temperatures should not exceed 60°F. Fertilization is stopped. It takes six to eight weeks for the flower buds to initiate.

The plants can be forced to flower after small buds become apparent. Forcing takes four to six weeks. Night temperatures are raised to 60 to 62°F. Feeding is resumed with 200 to 250 ppm nitrogen and potassium.

REVIEWING

MAIN IDEAS

Flowering potted plants are grown for their showy flowers. As a whole, they rank second in sales only to bedding plants. Some of the major potted flowering crops include poinsettias, chrysanthemums, African violets, and florist azalea. Some minor crops include kalanchoe, cyclamen, gloxinia, holiday cactus, florist hydrangea, Persian violet, cineraria, pocketbook plant, and primrose.

The crops are propagated by seeds or cuttings. The growing media are selected based on what each crop prefers. The same is true with the selection of growing pots. In general, vigorous vegetative growth is encouraged before the initiation of flower buds.

Each flowering potted crop can be brought to flower by manipulating the greenhouse environment. Flowering potted plants are often photoperiodic and thermoperiodic. Therefore, growers adjust the day length and temperatures to promote certain plant responses. Watering practices and fertilizer rates are also adjusted to meet scheduling needs. Light intensity is increased at times with supplemental lighting and decreased with shade curtains.

QUESTIONS

Answer the following questions using complete sentences and correct spelling.

1. What general conditions are given to promote vegetative growth of potted flowering crops?
2. Why are various crops pinched?
3. How are chrysanthemum varieties classified?
4. What does it mean to have a chrysanthemum with a nine-week response?
5. Why are chrysanthemums disbudded?
6. What are the general growing requirements for African violets?
7. Why are cineraria, pocketbook plant, and primrose often grown together?

EVALUATING

Match the term with the correct definition. Write the letter of the term on the line provided.

a. breaks
b. center bud removal
c. kalanchoe
d. heat delay
e. soft pinch
f. florist azalea
g. cyclamen
h. disbudding
i. potted flowering plants
j. pocketbook plant
k. liners
l. Persian violet

______1. The removal of 1/2- to 3/4-inch of the stem.

______2. Branching of the plants caused by pinching.

______3. Removal of flower buds to improve the overall quality of the plants.

______4. Plants grown in pots for their showy flowers.

______5. A cool crop from the area from Mexico to Chile that has colorful flowers resembling a woman's purse.

______6. The pinching off of the most terminal flower bud.

______7. Rooted cuttings pinched four to five times to encourage branching and sold to growers.

______8. A warm season plant with violet-blue flowers grown for Mother's day through October sales.

______9. Woody plants requiring between two and three years from propagation to flower.

_____10. Delay of flowering caused by warm temperatures.

_____11. Succulent plants with thick, fleshy leaves native to Africa and Asia.

_____12. A cool crop native to the Mediterranean with unusual-shaped white, pink, carmine, and red flowers.

EXPLORING

1. Grow potted flowering crops in your school greenhouse. Determine when you want the crop to flower. Then, schedule the crop from propagation to sale. Purchase seed, plugs, or cuttings. Plant them in the recommended growing medium and in the appropriate pots. Provide them with the recommended light duration, light intensity, temperatures, watering practices, and fertilization rates. Pinch the plants if necessary. Sell the finished crop to students, school staff, and/or the community.

2. Arrange a job-shadowing program with the assistance of your instructor, counselor, school administrators, and local greenhouse grower. Visit the greenhouse at least six times during the year. During the visits, observe the production practices employed by the grower. Also, consider whether greenhouse production is the type of career you would like to pursue.

CHAPTER 11

Growing and Forcing Bulbs, Corms, and Tubers

OBJECTIVES

This chapter is about growing and forcing bulbs, corms, and tubers. It has the following objectives:

1. Identify major crops grown from bulbs, corms, and tubers.
2. Differentiate between bulbs, corms, tubers, and rhizomes.
3. Explain the major steps in Easter lily production.
4. Describe cultural practices in forcing tulip crops.
5. Describe production practices associated with daffodils.
6. Explain how crocuses, dwarf irises, and grape-hyacinths are produced for sale.
7. Describe production practices associated with Asiatic and Oriental lilies.
8. Discuss how amaryllis crops are produced.

TERMS

amaryllis
Asiatic lilies
basal plate
bulb
case cooled by forcer
case cooled by supplier
controlled temperature forcing (CTF)
crocus
daffodil (narcissus)
double nose (DN)
dwarf iris
Easter lilies
forcing
geophyte
grape-hyacinth
hyacinth
leaf counting
natural cooling
non-precooled
non-tunicate bulbs
Oriental lily
precooled
rhizome
rooting room A
rooting room B
scales
scalettes
splitting
tuber
tulip
tunic
tunicate bulb

11–1. Bulb crops are forced for winter sales.

SPRINGTIME! For many people, it doesn't come soon enough. Winter days seem to last much too long. The outdoor landscape is not as green and colorful as it is during the other seasons. Plants help people to brighten their homes.

If there is one floriculture crop that carries the message of springtime, it is the bulb crop. Tulips and daffodils are associated with the arrival of spring and warmer temperatures to come. They are among the first plants to emerge from winter's sleep to flower in the landscape. Easter lilies, used widely in churches, are another signal of the beginning of spring. Forced to flower in the winter and in the early months of spring, bulb crops uplift the spirits of people tired of winter.

BULBS, CORMS, TUBERS, AND RHIZOMES

Potted bulb crops are grown for winter and spring sales. Some common bulb crops include Easter lily, tulip, narcissus, and hyacinth. Dwarf irises, crocuses, grape-hyacinths, Asiatic lilies, Oriental lilies, and amaryllises are other bulbs. Although these plants are commonly referred to as bulb crops, they technically are not all bulbs.

BULBS

Bulbs are short, flattened stems that bear fleshy, food-storage leaves. The fleshy modified leaves, called ***scales***, store food and water. At the base of each scale, there is a bud. The term used for plants that have their buds under ground is ***geophyte***. The scales can be relatively thin and wrap tightly around the bud, as in the case of tulips and daffodils, or they can be thick and loose, as with lilies.

11–2. Easter lilies have thick, loose scales and are non-tunicate.

11–3. This amaryllis has a dry, papery covering called a tunic. (Courtesy, DeVroomen Holland Garden Products)

True bulbs are further classified as tunicate or non-tunicate. A ***tunicate bulb*** has a dry, papery covering called a ***tunic***. Just as a coat protects you from the elements, a tunic protects the bud scales from damage and drying. Examples of tunicate bulbs include tulips, hyacinths, and narcissus. ***Non-tunicate bulbs*** lack a tunic and the protection a dry, papery covering would provide. Lilies are non-tunicate bulbs.

The bulb scales are held together at the bottom of the bulb by the basal plate. The ***basal plate*** is a hardened portion of stem tissue. Roots develop along the outside edge of the basal plate.

11–4. Shown are crocus corms. (Courtesy, International Flower Bulb Centre)

CORMS

Corms, although they look like bulbs, differ in structure. A corm is defined as a short, swollen, underground stem. It lacks fleshy scales. One or two dry leaf bases similar to a tunic cover corms. As with bulbs, new roots grow from a basal plate. Each season of growth depletes all the food and water reserves in a corm. New corms develop from buds located on the top or side of the original corm. Crocuses, freesias, and gladiolas are corms.

TUBERS

A ***tuber*** is an underground stem. It differs from bulbs and corms in that it has no dry leaf coverings or a basal plate. Tubers are usually flat and rounded. They have a tough outer skin from which roots develop. Tuberous plants include tuberous begonias, gloxinias, and anemones.

11–5. Tuberous begonias have tubers.

11–6. Calla lilies have rhizomes. (Courtesy, International Flower Bulb Centre)

RHIZOMES

A ***rhizome*** is an underground horizontal stem. Buds are produced on the top or sides of the rhizome. Irises and calla lilies are rhizomes.

BULB CROPS

Bulbs and corms are produced in regions of the world that provide optimum growing conditions. Holland produces the greatest number of bulbs. Some other areas that produce bulbs include Israel, the Pacific Northwest, and Michigan. Parts of the United States that have a short spring followed quickly by hot summer temperatures are not good for bulb growth.

Summer is a very important time in bulb production. A climate that provides long, cool, bright days is best for most bulbs. These conditions are good for vegetative growth. Leaves that stay on a plant for a long period produce sugars that contribute to the enlargement of the bulbs. Food produced is stored in the bulb. Bulbs are never dormant. They continue to grow and develop throughout the year, even when stored at near freezing temperatures!

It is best to pot and root bulbs before they are cooled. Roots that have developed are present to absorb water once the bulbs are moved to the greenhouse for forcing. As a result, rooted bulbs are well prepared to send all their stored energy into the development of leaves and flowers.

Bulbs require a vernalization, or cold period, to flower. Tulips, daffodils, lilies, crocuses, and irises are ***precooled***, or given cold treatment, for early forcing. Lilies are the only bulbs that might not be given some cooling prior to shipment. Bulbs that are not given cold treatment by the supplier are said to be ***non-precooled***. Precooled lilies have received all their cooling. Once bulbs have received the required cold period, they are forced. ***Forcing*** is a term used to describe the practices that get bulbs to grow and produce flowers.

Table 11–1. General Bulb Growth and Development Through the Year

	Programming of Crop			Forcing
Season	**Summer**	**Fall**	**Winter**	**Spring**
Growth and Development Stage	—Harvest of bulbs —Preplant storage for flower development	—Pot bulbs and root under cool, moist conditions	—Preparation for flowering under low temperatures —Temperatures lowered when roots grow out bottom of pot.	—Given warm temperatures, flower stalk elongates and flower opens.

EASTER LILIES (*Lilium longeflorum*)

General Information

Easter lilies are grown in the United States and Canada as a potted flowering crop for the Easter season. They produce a single stem and 1 to 20 large, fragrant, white, trumpet-shaped flowers. They are sold with one flower open. In Europe, Easter lilies are grown and sold as cut flowers. The primary varieties grown as potted plants are 'Nellie White' and 'Ace.'

The crop is difficult to grow because Easter falls on a different date each year. Easter is the first Sunday following a full moon after March 21. The earliest Easter can be is March 21. The latest date for Easter is around April 20. Therefore, growers must schedule their crops differently each year.

11–7. Easter lilies are grown as potted flowering plants in the United States and Canada. (Courtesy, International Flower Bulb Centre)

Propagation

Easter lilies are produced along the Pacific Coast in California and Oregon. The coastal area has cool, wet conditions all year. The bulbs are propagated from bulblets and scalettes. Bulblets form on the underground stem of a mature lily. ***Scalettes*** are small bulbs that grow on bulb scales that were removed from a bulb and planted.

In the first season, scalettes grow to 1 to 2 inches in circumference or the distance around the outside of the bulb. They are dug, graded, and replanted. A yearling is produced by the end of the second season that is 4 to 6 inches in circumference. They are dug, graded, and replanted. It takes three years to produce a commercial bulb with a 7- to 10-inch circumference. The harvested bulbs are packed in peat moss for shipment to growers around the country.

The forcer has a choice in what size bulbs to purchase. The larger the bulb, the more flowers it will produce. Bulbs are measured in circumference. They are graded and sold as $6^{1}/_{2}$ to 7, 7 to 8, 8 to 9, 9 to 10, and 10 to 11 inches. On average, a 10- to 11-inch bulb produces eight to nine flower buds on an 'Ace' and seven to eight flower buds on a 'Nellie White.'

11–8. These Easter lilies are undergoing natural cooling in a poly house.

Cooling

Easter lilies must go through vernalization. Vernalization is sometimes referred to as precooling or cooling. The required vernalization period for Easter lilies is 1,000 hours or around six weeks of cool temperatures. 'Nellie White' is cooled at 44 to 46°F and 'Ace' is cooled at 39 to 41°F. There are four methods of cooling Easter lily bulbs.

Controlled temperature forcing (CTF)—With this method, bulbs are received by the forcer and potted immediately. The growing medium is kept at 63°F for two to three weeks to promote the development of roots. Then, the bulbs are given 1,000 hours of 40°F temperatures. At the end of the cooling period, they are removed from the cooler and forced. CTF gives the forcer the most control over the cooling period. As a result, crops tend to be of higher quality. Plants produce more flowers and have longer leaves. The crops are easier to force and development is more uniform.

Natural cooling—This involves potting the bulbs as soon as they arrive. They are placed outdoors, or may be kept in a poly house if temperatures drop below freezing. Natural cooling is generally not recommended in middle or southern areas of the continental United States. As with controlled temperature forcing (CTF), cooling results in plants with more flowers and leaves.

Case cooled by supplier—The bulbs are cooled in their packing cases by the supplier before they are shipped to the customer. The forcer pots the bulbs upon arrival and begins the forcing process. This product is convenient for forcers that lack cooling space. A disadvantage is the cases at the center of the large coolers may not have been cooled long enough.

Case cooled by forcer—This is similar to case cooled by supplier with the exception that the cooling is done by the forcer. The advantage is the forcer can feel confident that the bulbs have received the required period of cooling.

Growing

Easter lilies are potted in 6-inch lily pots that are similar in size to a standard pot. Some growers plant two to three bulbs in larger pots. The growing medium should have a high bulk density and a pH between 6.5 and 7.0. Lilies are sensitive to fluorides. A symptom of fluoride damage common among monocot plants is leaf-tip burn. Sources of fluorides include perlite, city water, and super and treble phosphate fertilizers. Some growers avoid using perlite in their growing medium because of the fluoride.

The forcer must plant the lilies promptly upon receipt or, if the case is cooled by the forcer, after they have been cooled. Place the bulbs near the bottom of the pot. The bulb is

11–9. Although planted in the center, shoots often emerge at the edge of the pot. At this stage the plant on the right can be corrected.

then covered with several inches of growing medium. This potting practice promotes the formation of stem roots. The stem roots help to stabilize the plant and serve as a backup in case the bottom roots die.

Lilies should be watered sparingly at first. After the shoots have emerged, allow the medium to dry between waterings. Overhead watering is most commonly used with Easter lilies. Root rot can be a problem with wet medium, so capillary mats and ebb and flood systems are avoided. Fertilize lilies grown in soilless media with 250 ppm nitrogen and potassium. Medium with soil should receive 200 ppm nitrogen and potassium. Slow-release fertilizers are often incorporated in soilless mixes before potting. Early fertilization produces vigorous stem growth and leaf development.

Forcing

The simple schedule for CTF and naturally cooled bulbs is to pot the bulbs, keep them at 50 to 60°F for two to three weeks, cool, and force. Precooled or case cooled bulbs are potted, given one to two weeks at 50 to 60°F, and then forced. With six weeks of cooling, expect 110 to 115 days to force the plants.

11–10. These Easter lilies show excellent leaf development in the early stages of forcing.

Temperature affects the speed of forcing. It can be adjusted to speed or slow growth. If the lilies are too advanced, they can be given cooler temperatures to slow their metabolism. If the lilies are behind schedule, temperatures are raised. Generally, if Easter is early, bulbs are forced at 63 to 65°F nights. If Easter is late, the bulbs are forced at 60°F nights. When Easter is somewhere in the middle, they are forced at 62°F nights. Flower buds should be visible by Ash Wednesday or about 40 days before sale.

Long-day treatments can be used as a substitute for insufficient cooling or to induce earlier flower formation. Lighting plants from 10:00 p.m. to 2:00 a.m., beginning when shoots emerge and continuing for 10 to 14 days, speeds development. Each four-hour-per- day lighting substitutes one full day of cooling.

11–11. Temperature has a big effect on the speed of forcing Easter lilies.

The height of Easter lilies is influenced by a number of factors. 'Nellie White' is a shorter growing variety than 'Ace.' Light, temperature, humidity, and watering practices affect plant height. Lower light levels result in taller plants. The more water or fertilizer the plants receive, the taller the plant. Therefore, high temperatures that increase the frequency of water and fertilizer applications result in taller lilies. High humidity also produces taller plants.

11–12. Easter lilies can be grown fairly close together.

11–13. This grower counts leaves to determine the progress of his lilies.

DIF and growth retardants are employed to control the height of lilies. DIF is a very easy way to maintain the height of lilies. A zero or slightly negative DIF is the most affective. Large negative DIF causes the leaves to droop and curve down. Growth regulators, including A-rest, can be used as a drench or as a spray when shoots are 6 to 8 inches tall.

The practice of ***leaf counting*** is the best way to monitor the progress of the crop and to determine whether growth needs to be speeded or slowed. It also allows the timing of the crop to begin on January 15, leaving plenty of time for adjustments.

11–14. The grower holds a leaf identified by having its tip removed and counts the number of unfolded leaves to determine the rate at which the leaves are unfolding.

Table 11–2. Leaf Counting

Part 1—Determining the desired rate for leaves to unfold
1. Between January 15 and 20, select three to five plants to be the representatives for the crop.
2. Identify the uppermost unfolded leaf and mark it with a marking pen.
3. Count all of the leaves from the base of the plant up to the unfolded leaf. Record this figure. (Example: 46 leaves)
4. Remove and count all of the unfolded leaves. A magnifying lens and tweezers help with the smallest leaves. Record this figure. (Example: 48 unfolded leaves)
5. Compute the number of leaves that must unfold to meet the schedule. To do this, count backward from Easter to the date flower buds should be visible. Six weeks is the usual time period before Easter. (Example: If Easter is on April 4, flower buds should be visible February 21.)
6. There are 37 days from January 15 to February 21. Divide the total number of leaves yet to unfold by 37. The result is the number of leaves that must unfold per day in order to see flower buds by February 21. (Example: 48 divided by 37 = 1.3 leaves per day)
Part 2—Determining the rate of unfolding leaves
1. Select three to five plants to be the representatives for the crop. Place a label in the pots.
2. Identify the uppermost unfolded leaf and mark it with a making pen or notch it.
3. Wait four to five days and identify the new uppermost unfolded leaf and mark it with a marking pen or notch it.
4. Count the number of leaves that unfolded between the two marked leaves.
5. Calculate the number of leaves unfolding per day by dividing the number of new leaves unfolded by the days. (Example: Five newly unfolded leaves divided by four days = 1.25 leaves unfolding per day.)
6. Previously in the example, we determined that 1.3 leaves need to be unfolding per day to meet the schedule. The rate of 1.25 leaves that are actually unfolding is a little bit slow. Therefore, temperatures should be raised a bit to speed development.
7. By counting the leaves every four to five days, the rate of development can be monitored.

11–15. These lilies, with white, puffed-up flower buds, are ready for market.

Easter lilies are ready for market when the largest floral bud has ballooned and turned white. If necessary, plants that open early can be held in a cooler with temperatures between 35 and 40°F. They can tolerate cold storage for about two weeks. Warm them slowly when they are removed from storage.

If flower buds have opened, the anthers should be removed. Removal of the anthers prolongs the life of the flower. Also, the anthers produce an abundance of pollen. The pollen discolors the petals and can stain clothes and tablecloths.

11–16. These Easter lilies are being held in a cooler.

Pest and Disease Problems

Easter lily pests include the bulb mite. A recommended practice to control bulb mites is to dip the bulbs in a miticide bath before potting. Aphids and fungus gnats can also be problems.

Root rot is a concern for growers of Easter lilies. *Rhizoctonia* fungus can cause root rot especially in wet medium. Another fungus, *Botrytis*, can cause damage to flowers and flower buds. Virus infections can also deform flowers.

11–17. Before packaging for shipment, these people remove anthers from the flowers that have opened.

11–18. Bulb mites caused this lily to become deforme

SPRING FLOWERING BULBS

When spring flowering bulbs are mentioned, tulip, narcissus, hyacinth, dwarf iris, crocus, and grape-hyacinth come to mind. They require a cold treatment and force quickly. Spring flowering bulbs are considered to be high profit crops. They are relatively easy to force. They are in the greenhouse for a short period of time, typically from 5 days to 3 weeks. Only tulips need to be fertilized, and pesticides are seldom necessary.

Upon arrival, spring flowering bulbs should be unpacked and inspected for damage and disease. Store the bulbs in a well-ventilated area until planting. The conditions should be warm and dry, 63–64°F. Consider dipping the bulbs in a fungicide solution to control *Fusarium*, *Penicillium*, and *Botrytis*.

Use of shallow pots provides uniform cooling and greater control of watering in the cooler. Medium with sand or soil adds bulk density and can help keep bulbs from pushing up during the cooling stage. Avoid medium that contains perlite. Perlite contains fluorides that can cause leaf damage.

TULIPS (*Tulipa species*)

General Information

Most ***tulip*** species are from southwestern Europe and Asia Minor. They are very popular bulb plants. Tulips are available in practically all colors with the exception of a true blue and are grown for cut flowers or as potted flowering plants. The season for tulip production extends from early January to mid-May.

Tulips may be purchased as precooled or non-precooled bulbs. Precooled tulips require only 10 to 12 weeks of cold temperatures. For later forcing, approximately 18 to 20 cold weeks are needed. It is recommended that bulbs, 12/14 centimeters in circumference, be used for potted crops. Plan on 3 bulbs in a 4-inch pot, 4 bulbs in a 5-inch pot, 6 to 7 bulbs in a 6-inch pot, and 9 to 10 bulbs in an 8-inch pot. Use a well-drained medium with a pH between 6.0 and 7.0.

11–19. Potted tulips are very popular. (Courtesy, International Flower Bulb Centre)

Table 11–3. Number of Tulip Bulbs Per Pot

Container Size	Bulbs Per Pot
4-inch	3
5-inch	4
6-inch	6–7
8-inch	9–10

11–20. These bulbs are being grown in 8-inch pots.

11–21. Note how the first big leaf unfolds outward.

Growing

Plant the bulbs upon arrival. In the planting process, remove any bulbils attached to the bulb and remove the tunic. Place the bulbs with the flat sides facing the rim of the pot. By doing so, the big leaf develops to the outside of the pot. The nose or tip of the bulb should be slightly below the surface of the medium.

Tulip production is divided into two groups based on the forcing program. One group called ***rooting room A*** involves production for early flowering, before February 14. The other group, ***rooting room B***, involves production for later flowering, after February 14. Bulbs for rooting room A should be received and planted in mid-September. Order bulbs for a rooting room B program for planting around the beginning of October. Most tulip varieties require 15 to 16 weeks of cold temperatures followed by three to four weeks in the greenhouse to force them.

Rooting Room

After potting, the bulbs are cooled. This is often accomplished in poly houses or in refrigerated coolers. In the rooting room, bulbs are given 48°F temperatures for four to six weeks, or until roots begin to grow out the bottom of the pots. In this phase, the soil should be kept moist and the humidity maintained between 95 and 100 percent. The temperature is lowered to 41°F, and maintained until the shoots are about 1 inch high. At this stage, temperatures are dropped again to 31 to 35°F for the duration of the cold-week requirement.

Table 11–4. Rooting Room Temperatures and Time

Rooting Room	Flowering Schedule	Treatment
Rooting room A—Early	Early to middle season varieties flowering until February 14	48°F for 4–6 weeks (Nov. 10), 41°F until shoots 1 inch (Jan. 5), 32–35°F to end of cold requirement
Rooting room B—Later	Middle to late season varieties flowering after February 14	48°F for 4–6 weeks (Dec. 5), 41°F until shoots 1 inch (Jan. 5), 32–35°F to end of cold requirement

Forcing

In the greenhouse, strive for constant temperatures. Normally, 60°F is good. Provide good ventilation and keep humidity levels low. Tulips force best with light intensity between 1,000 and 2,500 foot candles. Keep the medium moist and water in the morning to reduce possible damage from *Botrytis*. Fertilize with 300 ppm of calcium nitrate.

11–22. These tulips have just been removed from the cooler and are ready for forcing.

11–23. This photograph shows four groups of tulips scheduled one week apart in order to provide a steady supply to retailers.

Move the tulips to market when they are in the bud stage, when at least one bud is showing color. The preferred height is 10 to 14 inches. If necessary, tulips can be stored in the green bud stage at temperatures between 33 and 35°F.

11–24. The flowers of large and miniature daffodils brighten the interiors of many homes.

DAFFODILS (Narcissus species)

General Information

Daffodils are very popular bulbs. The typical daffodil brightens any setting with its brilliant yellow flowers. Some varieties are available in white, two-tone colors, and peach. Potted daffodils are grown for sale during the months of January through April. Daffodils are native to Spain, Portugal, and Southern France.

Use only ***double nose (DN)*** bulbs when growing potted daffodils. DN I bulbs produce more flowers than DN II bulbs and DN II bulbs produce more than DN III bulbs. The number of bulbs planted in a pot depends on the bulb size and the size of the pots.

Growing

Pot daffodil bulbs upon receipt. Use well-drained medium with a pH of 6.0 to 7.0. Plant the bulbs at a depth whereby the widest part of the bulb is below the surface, but the nose is uncovered. Water thoroughly and move to the rooting room.

Table 11–5. Recommended Pot Size andNumber of Bulbs Per Pot

Pot size	4-inch pot	5-inch pot	6-inch pot	7-inch pot
DN I	Not recommended	Not recommended	3	5
DN II	Not recommended	Not recommended	3	5
DN III (Mini-narcissus)	3	5	7	Not recommended

Rooting Room

Provide the potted bulbs with 48°F temperatures for four to six weeks. Keep the growing medium moist, but not wet during this time. When roots are seen growing out the bottom of the pots, reduce the temperature to 41°F until shoots are 1 inch high. Then, lower the temperature to 32 to 35°F until the cold requirement has been satisfied, about 8 to 13 weeks.

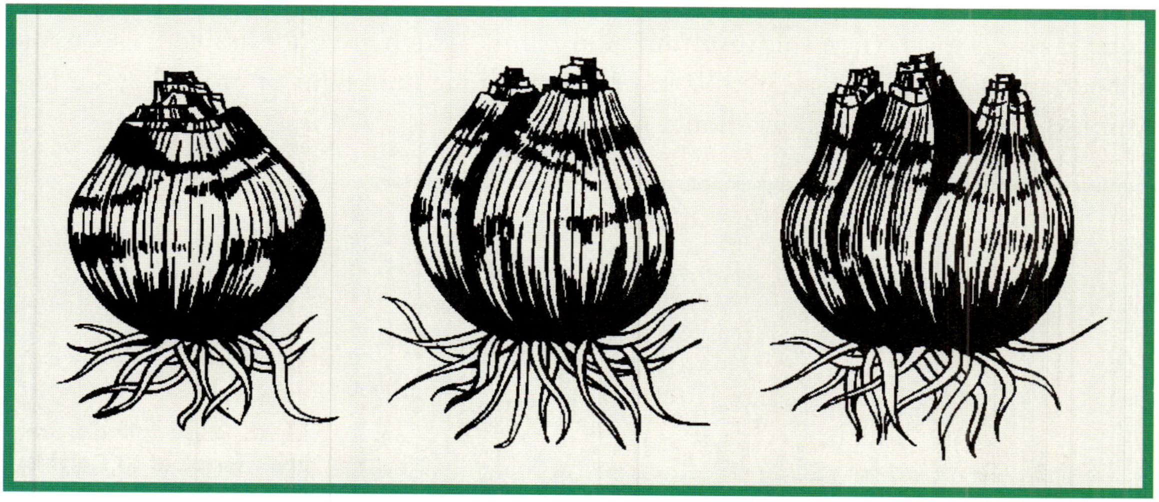

11–25. Pictured are a single, double, and triple nose daffodil bulb.

Each cultivar has an optimum number of cold weeks for a specific flowering date. Most varieties require 15 to 16 weeks of cold temperatures. Daffodils that have been precooled by

11–26. Three bulbs are placed in a 6-inch pot with the noses of the bulbs above the growing medium.

11–27. When shoots are 1 to 1½ inches in length, temperatures are lowered to 32 to 35°F.

the supplier require a total of 9 to 10 weeks of cold temperatures. Overcooling results in tall plants that are of less value.

Forcing

Move the plants to the greenhouse and force them at 60 to 63°F night temperatures. Temperatures lower than this result in taller plants, which are undesirable. Also, avoid temperatures above 65°F. Daffodils force best with medium light or around 2,500 foot candles. The growing medium should be kept moist.

11–28. These daffodils are being forced at 63°F nights.

Forcing of an early crop takes about three to four weeks at 63°F nights. Later crops require only one to two weeks in the greenhouse. These times vary slightly with the cultivar being grown. Although diseases and insects are seldom a problem, the forcer should watch for *Botrytis* and aphid problems.

Move the crops to market when the plants are in the pencil stage. This is before the flowers show color and enter the bent, "goose-neck," stage. A plant height of 8 to 12 inches is preferable. If necessary, store daffodils at temperatures of 33 to 35°F.

11–29. These paperwhite narcissus are nearly ready for market.

HYACINTH (*Hyacinthus orientalis*)

General Information

Hyacinths are native to Greece and Asia Minor. They produce a fragrant flower that may be purple, white, pink, or multicolored. Most hyacinths are produced in Holland. The majority of hyacinths are sold between January 3 and Valentine's Day.

Hyacinths are available to forcers as prepared bulbs or regular bulbs. Prepared bulbs require about 10 weeks of cold. They are used for early forcing. Regular bulbs need about 13 weeks to satisfy the cold requirement. Regular bulbs produce a better-looking plant than do prepared bulbs.

11–30. Hyacinths are fragrant as well as colorful.

Hyacinths come in sizes ranging from 15/16 centimeters in circumference to 19/20 centimeters. They require a very well-drained growing medium. They can be grown with one bulb in a 4-inch pot or a number of bulbs in larger pots. Use of short bulb pans provides a stable appearance.

11–31. Shown are seven hyacinths in an 8-inch bulb pan.

Count backward on a calendar the number of weeks to produce the crop, beginning from the sale date. The bulbs might need to be held from the time of their arrival until cooling. Hold prepared bulbs at 48 to 50°F and regular bulbs at 63°F if necessary.

Rooting Room

Provide rooting room temperatures between 48 and 50°F for about four weeks or until roots grow out the bottom of the containers. Drop temperatures to 41°F until shoots are 1 to $1^1/_2$ inches high. Lower temperatures to 33 to 35°F until ready to force.

Forcing

The longer the cooling period, the shorter period of time it takes to force hyacinths. December crops can be forced at 73°F greenhouse temperatures. Force January and February crops at 65°F. Later crops are forced at 60°F. Hyacinths require only a few weeks in the greenhouse. Sell the crop before it flowers, and let the buyer force the plant the last three to four days.

11–32. These hyacinths are nearly ready for sale.

One problem of hyacinths is known as splitting. ***Splitting*** is a condition where the flower stalk separates from the basal plate. Once separated, the flowers fail to develop. It is caused by changes in temperature that cause the bulb to expand and contract. Freezing of the bulbs can also lead to this problem. Splitting is more common with purple varieties.

CROCUS (*Crocus* species), DWARF IRIS (*Iris reticulata*), and GRAPE-HYACINTH (*Muscari armeniacum*)

General Information

Crocuses are a very popular small plant native to central and Western Europe. Crocus varieties offer flowers of bright purple, yellow, and white colors. **Dwarf irises** are native to the Caucasus region and are purple in color. **Grape-hyacinths** are native to Asia Minor. Grape-hyacinths have clusters of blue or white flowers. The blues can sometimes appear iridescent. The majority of bulbs for these three crops are produced in Holland.

11–33. Crocus (left), dwarf iris (center), and grape-hyacinth (right) all have similar forcing schedules. (Courtesy, International Flower Bulb Centre)

The recommended bulb sizes for crocus and grape-hyacinth are 9 cm and larger. Bulb size recommendations for dwarf iris are 6 cm and larger. The bulbs are grown in 4-, 5-, and 6-inch containers. A well-drained growing medium with a pH between 6.0 and 7.0 is suggested. Plant the bulbs immediately upon arrival. The proper planting depth involves a slight covering over the tops of the bulbs.

Table 11–6. Suggested Number of Bulbs Per Pot

	4-inch pot	5-inch pot	6-inch pot
Crocus	5–6 bulbs	7–9 bulbs	10–12 bulbs
Dwarf iris	7–10 bulbs	10–12 bulbs	12–15 bulbs
Grape-hyacinth	5–6 bulbs	7–9 bulbs	10–12 bulbs

Rooting Room

Provide 48°F temperatures for four to six weeks or until roots emerge from the bottoms of the pots. Reduce the temperature to 41°F, and store at this temperature until the shoots are about 1 inch high. Then, lower temperatures to between 32 and 35°F until the completion of the cold-week requirement.

The cold-week requirement for these crops is 15 to 16 total weeks. Precooled bulbs need only 9 to 10 weeks of cold before forcing. During the rooting room treatment, keep the growing medium moist.

11–34. Dwarf irises force very quickly from this stage.

11–35. The floral sheaths of these crocus are visible making them ready for market.

Forcing

Crocuses and dwarf irises force very quickly. For early forcing, they develop in only one to two weeks in the greenhouse. Later forcing takes only two to three days. Grape-hyacinth, on the other hand, needs about three weeks in the greenhouse for early forcing. Later forcing takes about two weeks.

Provide temperatures of 55 to 60°F nights and 60 to 63°F days. Light requirements for forcing are considered low to medium, 1,000 to 2,500 foot candles. For early forcing, induce the shoots to stretch by keeping them dark for a few days.

Market crocuses and dwarf irises as the sheaths encasing the flower buds become visible. Late-season crops can go right from the rooting room to market. Since these two plants have short shelf lives, it is appropriate to get them to the customer early in their flowering stage. Grape-hyacinth should be marketed when the flower buds show the first hints of color. If necessary, short-term storage can be accomplished with temperatures of 33 to 35°F.

ASIATIC AND ORIENTAL LILIES (Lilium hybrids)

General Information

Asiatic lilies and **Oriental lilies** are, as implied by their name, native to the Asian Continent. They come in a wide range of colors including yellow, white, red, and orange. They are forced throughout the year.

11–36. Asiatic lilies (left) and Oriental lilies (right) come in a wide range of colors. (Courtesy, International Flower Bulb Centre)

Most Asiatic lilies and Oriental lilies are precooled by the supplier. Therefore, no additional cooling is required by the forcer. Asiatic lilies are precooled for six weeks and Orientals for 8 to 10 weeks at 34 to 35°F.

Growing

Pot Asiatic lilies in deep pots. A common practice is to place three bulbs in a 6-inch standard pot and five bulbs in an 8-inch standard pot. The growing medium should be well drained and have a pH within a range of 6.0 and 6.5.

11–37. Amaryllis produce beautiful, large flowers. (Courtesy, DeVroomen Holland Garden Products)

Grow Asiatic lilies at 55 to 60°F nights and 70°F days. Orientals prefer 65 to 67°F nights and 75°F days. Temperatures control timing of the crops. Also, DIF can be used to control the height of the plants. The length of time from planting to flowering varies with the variety from 56 days to 106 days. They are ready for market when the first is swollen and shows color.

AMARYLLIS (*Amaryllis belladonna*)

General Information

Amaryllises can be found growing naturally in South America. Varieties of amaryllises have been improved through hybridization providing a wide range of large, colorful flowers. Amaryllises are typical of subtropical plants. As such, they are not cooled, nor do they like cold temperatures. They are actually grown more like a houseplant. Bulbs are produced in Israel, South Africa, and Holland.

The majority of amaryllis cultivars produce four flowers on a stalk. Larger bulbs often produce two flower stalks. Bulbs are available as 20/22, 24/26, 28/30, and 32/up centimeters in circumference.

11–38. This amaryllis is being prepared for repotting.

Growing

Pot amaryllis bulbs as soon as they arrive from the supplier. Use pots slightly larger than the bulbs. Six-inch pots work well with most bulbs. In the process, clean the bulbs of debris and dead leaves, but leave healthy roots undamaged. Place the bulb at a level that will leave its shoulders exposed. A growing medium with a pH of 6.0 to 6.5 is preferred.

Water thoroughly at first, then sparingly for a few weeks. Once the leaves develop, water regularly. Fertilize with a complete fertilizer every two to four weeks. Grow the plant at temperatures from 70 to 80°F. An amaryllis should receive medium light intensity, 2,500 to 5,000 foot candles.

An amaryllis begins to flower about eight weeks after potting. Plants are marketed when a floral stalk is about 12 inches high. It is also desirable to have leaves unfolding and 6 to 12 inches in length. If development needs to be slowed, store them at 48°F. Temperatures lower than 48°F can cause damage.

Reforcing

In September, stop watering the plant. If possible, give the plant 50 to 60°F temperatures. The leaves will dry and the bulb will go dormant. Allow 10 to 12 weeks for this downtime. In November/December clean the bulb and repot.

PLANTING OUTDOORS

After flowering, bulb crops can be planted outdoors. Of course, bulbs that require vernalization do well in the northern parts of the country where they receive natural cooling. Tender crops, such as tuberous begonia and amaryllis, do well in the warm areas. Plant the bulbs at the recommended depth. Placement of bone meal fertilizer in the planting hole boosts nutrient levels and promotes plant growth.

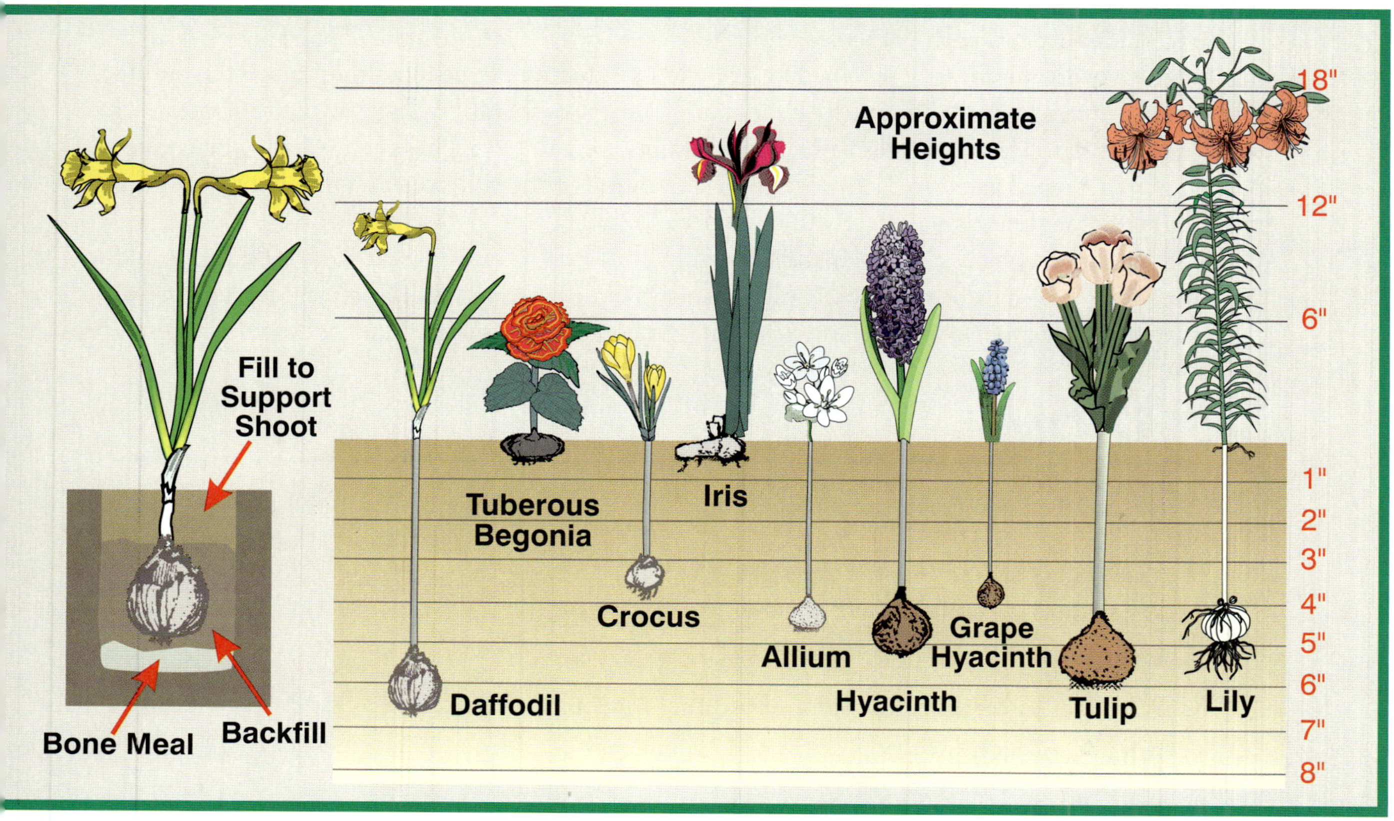

11–39. Planting depths for select bulbs.

REVIEWING

MAIN IDEAS

Bulb crops include those plants that have bulbs, corms, tubers, and fleshy rhizomes. The major crops are Easter lilies, tulips, narcissus, crocuses, dwarf irises, grape-hyacinths, Asiatic lilies, oriental lilies, and amaryllises. True bulbs can be further divided as being tunicated or non-tunicated. Bulb crops are typically grown for late winter and spring sales.

A great number of bulbs are produced in Holland, the Pacific Northwest, and Michigan. Long, cool, bright days are ideal for promoting vegetative growth and bulb expansion. Bulbs require a vernalization or cold period to flower. Tulips, daffodils, lilies, crocuses, and irises are precooled. All but the lilies are given some cooling prior to shipment. Bulbs that are not given cold treatment by the supplier are said to be non-precooled.

Easter lilies are grown in the United States and Canada as a potted flowering crop for the Easter season. The primary varieties grown as potted plants are 'Nellie White' and 'Ace.' The bulbs are propagated from bulblets and scalettes. It takes three years to produce a 7- to10-inch circumference bulb. Easter lilies must go through vernalization, referred to as precooling or cooling. Temperature affects the speed of forcing and can be adjusted to speed or slow growth. The rate of growth can be determined through a practice known as leaf counting.

Tulips, daffodils, hyacinths, crocuses, dwarf irises, and grape-hyacinths are popular spring flowering bulbs. They are purchased precooled for pot plant production. In general, they are given rooting room temperatures between 48 and 50°F for about four weeks or until roots grow out the bottom of the containers. Drop temperatures to 41°F until shoots are 1 to $1^1/_2$ inches high. At this stage, temperatures are dropped again to 31 to 35°F for the duration of the cold week requirement. They are moved to the greenhouse and forced at 60 to 63°F.

The supplier precools Asiatic lilies and Oriental lilies. Grow Asiatic lilies at 55 to 60°F nights and 70°F days and Orientals at 65 to 67°F nights and 75°F days. Amaryllis are typical of subtropical plants and are grown like a houseplant. They are not cooled, nor do they like cold temperatures.

QUESTIONS

Answer the following questions using complete sentences and correct spelling.

1. How do bulbs, corms, tubers, and rhizomes differ?
2. What does a tunicate bulb mean?
3. Why are bulbs precooled?
4. Why are Easter lilies considered a difficult crop to grow?
5. What is vernalization?
6. Why is leaf counting practiced?
7. When are rooting room A and rooting room B used?

8. What is the purpose of the rooting room?
9. Why is it important to get crocus and dwarf iris to the customer quickly?
10. How does an amaryllis differ from the spring bulbs?

EVALUATING

Match the term with the correct definition. Write the letter of the term on the line provided.

a. hyacinth
b. non-tunicate bulbs
c. non-precooled
d. splitting
e. amaryllis
f. case cooled by supplier
g. corm
h. forcing
i. scalettes
j. leaf counting

______1. Lack a tunic and the protection a dry, papery covering would provide.

______2. Bulbs are cooled in their packing cases by the supplier before they are shipped to the customer.

______3. Defined as a short, swollen, underground stem.

______4. Typical of subtropical plants, they are not cooled.

______5. Small bulbs that grow on bulb scales that were removed from a bulb and planted.

______6. Bulbs that are not given cold treatment by the supplier.

______7. A condition where the flower stalk separates from the basal plate.

______8. Practice used to monitor the progress of an Easter lily crop and to determine whether growth needs to be speeded or slowed.

______9. A term used to describe practices that get bulbs to grow and produce flowers.

_____10. Produce a fragrant flower that may be purple, white, pink, or multicolored.

EXPLORING

1. Obtain tulips, narcissus, hyacinths, crocuses, dwarf irises, or grape-hyacinths in the fall of the year. Pot them as instructed in the text. Place them in a refrigerator set at the optimum temperatures for the recommended period of cold time. Remove the bulbs and force them in the home or in a greenhouse.

2. Continue to grow potted bulbs from late winter and early spring indoors until the weather warms up enough to safely plant them outside. Plant the bulbs with bone meal at the correct depth. Wait for next spring to see flowers.

3. Propagate your own amaryllis. As the bulbs grow year after year they produce side bulbs. Remove these when you repot the amaryllis and pot them in smaller pots.

CHAPTER 12

Growing Cut Flowers and Foliage

OBJECTIVES

This chapter is about growing cut flowers and foliage. It has the following objectives:

1. Recognize the importance of the international and domestic cut flower markets.
2. Identify major cut flowers.
3. Describe cultural practices in growing cut flowers.
4. Recognize pest and disease problems of major cut flower crops.

TERMS

alstroemeria
freesia
gerbera daisy
gladiolus
hook cut
hybrid tea rose
leaf drop
mini carnation
single flowering carnation
snapdragon
splitting of the calyx
standard carnation
sweetheart (floribunda) rose

12–1. Roses are the most recognized cut flower and are easily purchased as a bouquet. (Courtesy, Steven E. Newman, Colorado State University)

ROSES are red. Violets are blue. Who hasn't made up a poem beginning with those lines? The familiarity of that poem underscores how roses have a permanent place in American traditions and way of life. People value roses as cut flowers. Cut flowers, easily purchased as a bouquet, can be placed in a vase and brighten a serving table at a party. Gifts of cut flowers can be given to express love, sympathy, and as a request for forgiveness.

Although roses might be the most recognized cut flower in the United States, there are many types of flowers grown for use as cut flowers. As production techniques, marketing, and distribution of cut flowers have improved, consumers have been given greater variety. That variety is evident at supermarkets that package cut flowers for cash and carry sales. Florists also have an increased selection of flowers for use in all kinds of design work.

CUT FLOWER OVERVIEW

Times have changed! Before World War II, floriculture production in the United States meant cut flowers. If your grandmother or great-grandmother were to go to the market to buy flowers, they would have been shopping for cut flowers. Potted flowering plants, foliage plants, and bedding plants were not common then. Now, cut flower and cut foliage production in the United States is valued at less than 12 percent of the entire floriculture production.

According to the United States Department of Agriculture, production of cut flowers and foliage in the United States amounted to $535 million in 2001. That places cut flowers fourth behind the production of bedding plants, potted flowering plants, and foliage plants. Rose, lily, tulips, gladiola, and iris rank as the top cut flower crops in terms of wholesale value. Other crops include gerbera daisy, snapdragon, chrysanthemum, delphinium, lisianthus, orchid, alstroemeria, carnations, freesia, and exotic tropical flowers. The top five cut flower and cut foliage producing states include California, Florida, Washington, Hawaii, and Oregon. Hawaii is the primary producer of exotic, tropical cut flowers.

Table 12–1. Value of United States Cut Flower Crops Sold according to the 2002 USDA Floriculture Crops Summary

Crop	Wholesale Value in Millions of Dollars
All Roses	67,656
Lilies	57,452
Tulips	26,265
Gladiola	24,183
Iris	20,117
Gerbera Daisy	20,045
Snapdragons	17,249
Chrysanthemum, Pompon	16,578
Delphinium and Larkspur	10,807
Lisianthus	10,101
Orchids	8,883
Alstroemeria	5,206
Carnations, Standard	4,571
Other cut flowers (Miniature Carnations, Standard Chrysanthemum freesia, anthurium, etc.)	135,143

Florida is the largest producing state in terms of cut foliage or greens. Cut foliage, which includes leatherleaf fern, asparagus fern, pittosporum, and podocarpus, is used to add greenery to floral work. Florida held a 79 percent market share of domestic production in 2001. Other states with notable cut foliage production include California, Oregon, and Hawaii. Increasing amounts of cut foliage are being imported from Mexico, Costa Rica, Guatemala, and other countries.

Table 12–2. Value of United States Cut Foliage Crops Sold according to the 2002 USDA Floriculture Crops Summary

Crop	Wholesale Value in Millions of Dollars
Leatherleaf fern	54,283
Other cut cultivated greens (asparagus fern, pittosporum, podocarpus, etc.)	56,783

Table 12–3. Value of Hawaiian Tropical Crops Sold according to the 2002 USDA Floriculture Crops Summary

Crop	Unit	Wholesale Value in Millions of Dollars
Anthuriums	blooms	6,904
Birds of Paradise	stems	695
Pink Ginger	stems	492
Red Ginger	stems	755
Other Ginger	stems	169
Heliconia	stems	598
Orchids, Cymbidium	stems	232
Dendrobium Sprays	stems	2,990
Dendrobium Individual Blossoms	blooms	672
Vanda Individual Blossoms	blooms	340
Oncidium	stems	813
Protea	stems	1,122

The United States cut flower market relies heavily on imports from other countries. Imports of cut flowers in 2001 totaled $565 million, well exceeding the value of domestic cut flower production. South American exports to the United States accounted for $408 million of cut flowers shipped. The countries with major exports to the United States follow:

- Columbia, $302 million
- Ecuador, $99 million

12–2. Hawaii is the primary producer of exotic, tropical cut flowers including the bird of paradise. (Courtesy, Agricultural Research Service, USDA)

- Holland, $67 million
- Mexico, $29 million
- Canada, $18 million
- Costa Rica, $14.7 million

To remain profitable, growers in the United States have had to change. Some growers refocused their production to potted flowering crops and bedding plants. Some growers invested in technology for more efficient production. Others switched from growing major cut flower crops to specialty cut flowers that would not be in direct competition with imports.

Cut flower production is an international affair for a number of reasons. Modern air travel has made rapid shipments possible. Cut flowers, grown and harvested in one part of the world, can be transported to another part of the world in hours. Production of cut flowers is labor intensive. As a result, some crop production has shifted to parts of the world where the cost of labor is low. Also, in some parts of the world, such as Columbia, the temperatures, light, and other conditions are absolutely perfect for growing certain cut flowers.

PRODUCTION OF CUT FLOWERS

ROSE

Two types of roses are produced, hybrid tea roses and sweetheart (floribunda) roses. About 90 percent of the cut roses produced in the United States are hybrid tea roses. The ***hybrid tea rose*** is a large, usually solitary flower. The majority of the hybrid tea roses produced are red. Other popular colors include pink, yellow, and white. The ***sweetheart (floribunda) rose*** is a smaller rose that grows as a spray and is often disbudded. Sweetheart roses are used extensively in wedding work, corsages, and centerpieces. Yellow and pink make up a large percentage of those produced. New cultivars also offer many pastel colors.

12–3. The hybrid tea rose (left) is a large, usually solitary flower, and the sweetheart (floribunda) rose (right) is a smaller rose that grows as a spray.

Interestingly, most holidays are in sync with rose production. The holidays are spaced about six weeks apart during the year. It takes from six to eight weeks to produce a new crop of roses after a harvest. Eight weeks are required during the darker months of December. There is also a long break in production after June when rose plants are cut back. The two biggest holidays for rose sales are Valentine's Day and Mother's Day.

- Early September—Grandparent's Day
- October—Sweetest Day
- December—Christmas
- February—Valentine's Day
- April—Easter, Secretaries Day
- May—Mother's Day
- June—Weddings

Growing

Roses are an exacting crop. Temperatures must be maintained at optimum levels. In addition, pinching and lighting must be carefully controlled in order for growers to have the flowers ready for specific holidays. Given the right climate, cut roses can be grown successfully outdoors.

Propagation is done primarily by grafting. The most common procedure, t-budding, involves buds from select rose cultivars being placed on multiflora rose rootstocks. Ninety percent humidity is maintained in the greenhouse to help the grafts to take. Plants are planted with the bud union at or just below the soil surface.

Roses like an organic growing medium that is well drained. An organic matter content of 35 to 37 percent provides good drainage, aeration, and cation exchange capacity. The expected life of a rose bush in commercial operations is 5 to 7 years, so a medium with stable organic matter is important. A growing medium pH of 6.0 to 6.5 is preferred.

12–4. Roses are usually grown in organic growing medium. (Courtesy, Steven E. Newman, Colorado State University)

Fertilization requirements are considered medium to high. A suggested fertilization program involves the application of an ounce of granular 10-10-10 or similar fertilizer three times a year. The fertilizer is placed alongside the plant in April, June, and mid-August. A constant liquid feed applied through a drip irrigation system is another option. Feeding at a rate of 100 to 150 ppm nitrogen, phosphorus, and potassium is recommended. It is also recommended that every third watering be clear water.

Roses do best in warm temperatures. They like 62 to 65°F nights and 75 to 90°F days. Cooler temperatures result in larger flowers, but it takes them longer to produce. During periods of low light, the temperature in the greenhouse should be lowered. Roses also require full sun to maximize the production of sugars.

Pruning of roses every year, or at least every other year, helps maintain production levels. A direct prune using lopping shears is performed right after Mother's Day. With direct pruning, stems are cut back to within 18 to 30 inches of the rootstock. Gradual pruning can also be performed. It is a method used in the spring and fall, and it is usually done the first year or the first year after a direct cut. Gradual pruning involves the removal of a portion of the old wood with a flower. The cut is known as a **hook cut**.

Harvesting

Harvest roses when the sepals have dropped and the outer petals begin to open. Growers can expect hybrid tea rose plants to produce 20 to 30 flowers a year. Sweetheart rose plants produce around 40 stems a year. Where the cut is made on the stem influences both the stem length and the growth rate of the bush.

12–5. Harvest roses when the sepals have dropped and the outer petals begin to open. (Courtesy, Hills Floral Products, Inc., Richmond, Indiana)

Generally, the cut is made just above the node of the second five-leaflet leaf from the tip of the stem. Long stems are valued. However, when cuts are made below the second five-leaflet leaf, recovery of the bush and production of the next flowers are delayed. Stems cut 22 inches and longer in length are ordinarily used as gift roses. Those cut to the length of 10 to 14 inches tend to be used for floral work, such as bouquets.

Harvesting is performed in the morning and evening hours to reduce stress on the cut flower. Flowers are placed directly in water acidified with citric acid to a pH of 3.5. They are placed in a cooler set at 34°F. After a few hours, they are graded and placed in bunches of 12 to 25 stems for shipment. Grading classification is as follows:

- Short 10 to 14 inches

- Medium 14 to 18 inches
- Long 18 to 22 inches
- Extra long 22 to 26 inches
- Fancy 26 to 30 inches
- Extra fancy 30 or more inches

Pests and Diseases

One of the main problems with roses is ***leaf drop*** caused by variations in temperature, excessive drying, excessive soluble salts, or wet soil. Cool, moist conditions can lead to problems with black spot or mildew. Major pests include two-spotted mites, aphids, and rose midge.

12–6. Gerbera daisies come in many bright colors. (Courtesy, California Cut Flower Commission)

GERBERA DAISY

Gerbera daisy is also known as the Transvaal daisy. It is native to South Africa. Its large bright flowers come in white, yellow, pink, red, oranges, salmon, and violet. ***Gerbera daisies*** grown for cut flower production are most often propagated by tissue culture. They are transplanted to $2^1/_4$-inch pots. Newly potted plantlets are shaded and placed under mist. The mist is gradually reduced and after about 6 weeks plantlets are ready for planting.

Cut gerberas are grown in beds or benches. The growing medium must be sterile and well-drained. A pH of 6.5 is recommended. Care must be taken in planting gerbera daisies.

The crown or base of the plant must not be buried. Plant gerberas in two rows. The rows are spaced 14 to 18 inches apart, and plants within the rows 10 inches apart.

Overhead irrigation is used to ensure thorough wetting and to raise humidity around the plants. Once established, the plants are watered with drip irrigation. Gerberas have low fertilizer needs. Slow release fertilizers are used. Then, supplemental applications of 100–200 ppm N and K are applied with every watering.

Gerberas grow best under maximum light conditions. Day temperatures should range between 70 and 80°F. Optimal night temperatures should be around 60°F.

Gerberas are susceptible to a number of diseases including *Botrytis*, *Phytophthora*, *Pythium*, and *Fusarium*. Losses can be kept in check by keeping leaves dry and applying preventative fungicide drenches. Leafminers are the most common insect problem. Whiteflies and thrips also cause damage.

Flowers are harvested 8 to 12 weeks after planting. Individual stems should be showing two to three rows of stamens at the center of the flower. Stems are harvested by pulling them from the plant. The flowers should be placed in water immediately on a rack. The rack holds the flowers upright and reduces the curving of the stem.

12–7. A standard carnation (left) has a large solitary flower, and the mini carnation (right) has two to four smaller flowers on a stem.

CARNATION

Carnations are widely used in the floral industry, and rank as the third most popular cut flower in the world. Their long vase life and their tolerance to shipping increase their value. Colors range from red, white, pink, yellow, pastels, and variegated.

Three types of cut carnations are available. ***Standard carnations*** are large, solitary flowers that have been disbudded. Flowers of ***mini carnations*** are 1/2 to 1/3 the size of standard carnation flowers. The terminal flower bud of a mini carnation is usually removed to improve the development of a 2 to 4 flower spray. There are also single flowering types. ***Single flowering carnations*** produce a flower without the need of disbudding.

In the 1920s, 30s, and 40s, the Chicago area was considered the carnation capital of the world. Other areas in the United States known for carnation production include Southern California, the mountainous regions of North Carolina, and Colorado. Today, however, the United States imports most of its carnations. Columbia supplies roughly 80 percent of the carnations sold in the United States.

Carnations are propagated from stem cuttings. They are planted shallow in ground beds. The shallow planting helps to reduce verticillium, fusarium, and bacterial stem rot disease problems to which they are very susceptible. The medium must also be sterilized to reduce disease problems. The medium pH should be between 5.5 and 6.5.

12–8. Carnations are grown on a one- or two-year rotation usually in ground beds. (Courtesy, Steven E. Newman, Colorado State University)

Columbia, a mountainous country located near the equator, has perfect conditions for year-round carnation production. The natural temperatures are what carnations prefer, 50 to 55°F nights and 60 to 65°F days. The light intensity at higher elevations is excellent. Carnations are long-day plants and in Columbia, the day lengths are the same throughout the year. Production in the United States requires lighting from September through December. Also, growers in the United States must control greenhouse temperatures.

Carnations are grown on a one- or two-year rotation. Flowers in the first year are larger. There are more, but smaller, flowers produced in the second year. Standard carnations are harvested when about 1/2-inch of the petals show above the calyx. Mini carnations are harvested when two flowers are open and the buds are showing color. The biggest problem with harvested flowers is the ***splitting of the calyx***. This condition is caused by fluctuations of temperatures and is more common with some cultivars than others. When the calyx splits, petals spill out and the flower appears misshapen.

GLADIOLUS

Gladioli are valued for their large showy floral spikes. Cultivars are available in nearly every color. As cut flowers, they are long lasting. They can also be harvested when the first bud shows color. Yet unopened buds will open right on up to the top of the stem.

Much of the commercial production of gladioli takes place in Florida and California. However, gladioli corms are grown across the United States. In California, production is year-round, while November through June are the prime months of production in Florida.

12–9. Gladioli are available in many colors. (Courtesy, Joe Steffen)

Gladioli are produced outside in fields. Since they do not tolerate frost, planting in the north is not done until after the frost-free date. Typically, corms for cut flower production are planted about 1 inch apart in 5- to 6-inch deep furrows. Gladioli are relatively heavy feeders. Fertilization with a complete fertilizer at planting and side dressing during the development of the crop is recommended. After flowering and as the leaves die back, corms are dug and placed in dry storage.

Harvested gladioli are kept in an upright position. This is because they respond dramatically to gravitropism. Should they lie on their side for a short while, their stems bend upward. The resulting crooked stems are of less value to floral arrangers and designers.

12–10. This field of gladioli is used for hybridizing. (Courtesy, Joe Steffen)

Major pest problems include aphids and thrips. Thrips can be especially troublesome and cause damage to both foliage and flowers. Regular pesticide applications can be used to control thrips.

SNAPDRAGON

The elongated floral inflorescence of snapdragons is very attractive. **Snapdragons** come in a variety of bright colors. White, pink, and yellow are very popular. Other colors available include red, bronze, and purple. Children are often entertained by removing and pinching the sides of a floret to see the "dragon's" mouth open.

12–11. Snapdragons are very popular and come in a variety of colors. (Courtesy, Steven E. Newman, Colorado State University)

12–12. These snapdragons were grown from seed. (Courtesy, Steven E. Newman, Colorado State University)

Unlike most cut flower crops, snapdragons are propagated by seed. Seeds germinate in 5 to 7 days at 65 to 70°F. The seeds are left uncovered for best results. After they have developed the first set of true leaves, they are transplanted to the growing bed.

The hybrids used have been developed for specific periods of the year. Winter snapdragons have been bred to perform in lower light levels and cooler temperatures, 50°F nights. Spring snapdragons do well in high light and cool temperatures, 55°F nights. Summer snapdragons are bred for high light and warm temperatures. Fall snapdragons like high light and warm temperatures, 60°F nights.

Snapdragons are grown in ground beds or on raised benches. In either case, they like a well-aerated growing medium with a pH between 5.5 and 6.5. They are considered to be light feeders.

It is necessary to provide support for snapdragons with netting. Crops require two layers of netting with the top one at a level below the lowest flower. Keeping the flowers upright after harvesting, during storage, and while shipping is critical to maintain straight stems. Laid on their side, snapdragons will bend within half an hour.

The best-quality flowers are cut with a minimum of 5 to 7 florets open. The stems are placed in water as soon as possible. Foliage from the lower 1/3 of the stem is removed before the stems are graded and bunched.

Snapdragons have few serious pest or disease problems. Pest problems associated with snapdragon production include aphids, thrips, and whiteflies. Diseases of note are *Botrytis*, mildews, and damping off of seedlings caused by *Pythium*.

ALSTROEMERIA

Alstroemeria, also known as the Peruvian lily or Inca lily, has gained popularity in the United States since the 1970s. It is everblooming, produces high yields, and it is easy to grow under cool temperatures. Flowers of the *Alstroemeria* are borne on a cyme, and come in many colors including yellow, orange, red, pink, purple, lavender, white, and bicolor. Another asset is alstroemeria have a long vase life, two weeks or more.

Propagation of alstroemeria is done by division of its white fleshy rhizomes or by tissue culture. Plants are placed in ground beds since they can grow to a height of 6 feet. The medium should have a high percentage of organic matter. The pH should be between 6.0 and 6.5.

Alstroemeria requires high nutrient levels. During spring growth and flower production, they like 600 ppm of a 20-20-20 fertilizer on a weekly basis. They also require large amounts of water and high light intensity for maximum floral production. Alstroemerias are long-day plants and, as such, require lighting during the darker months to promote flowering.

Flower clusters are harvested when the first flower begins to open. Pests and disease problems are minimal. Aphids, mites, and *Botrytis* are the chief problems.

12–13. Alstroemeria are long-lasting flowers that have increased in popularity. (Courtesy, Steven E. Newman, Colorado State University)

12–14. Alstroemaria are long-day plants that require large amounts of nutrients and water. (Courtesy, Steven E. Newman, Colorado State University)

12–15. Freesias are fragrant flowers that come in bright colors. (Courtesy, Steven E. Newman, Colorado State University)

FREESIA

Freesia is a cut flower that has grown rapidly in popularity. The lily-like flowers, borne on a horizontal spike, come in bright colors including yellow, red, white, blue, and pink. In addition to the visual interest, they are very fragrant. Some people liken their scent to the breakfast cereal, Fruit Loops. Freesias are produced from corms.

Freesias are planted in September in northern regions and as late as December in southern regions of the United States. Corms are planted 2 inches deep in ground beds or raised beds at a spacing of about 12 corms per square foot. Freesia requires a well-drained medium with a pH between 6.5 and 7.2. Perlite must not be used in the medium, as freesias are sensitive to fluorides.

Grow freesias warm (60 to 65°F nights) after planting until the stems are about 12 inches in height. Then, drop the temperatures to 50 to 55°F nights to trigger flowering. Avoid letting the day

12–16. These freesias will be supported with wire and string supports as they grow. (Courtesy, Steven E. Newman, Colorado State University)

temperatures exceed 63°F after the temperatures have been dropped. Freesias like high light intensity and 200 ppm of a 20-20-20 fertilizer every other week. The first flowers appear about eight weeks following the onset of cool temperatures. Each stem produces approximately six flower stalks in the course of a season. Harvest flowers when the first floret opens.

Thrips and aphids must be controlled as they can serve as vectors for viral disease. Other problems include fusarium and spider mites.

REVIEWING

MAIN IDEAS

The floriculture industry has changed. Cut flower production, once the most important aspect of the floriculture industry in the United States, now makes up less than 17 percent of the entire floriculture production. That places cut flowers fourth behind the production of bedding plants, potted flowering plants, and foliage plants. The top five cut flower producing states include California, Florida, Hawaii, Colorado, and Pennsylvania.

The United States cut flower market relies heavily on imports from other countries. The countries with major exports to the United States include Columbia, Ecuador, Holland, Canada, and Mexico. Modern air travel has made rapid shipments possible. Production of cut flowers has shifted to parts of the world where the cost of labor is low. Also, in some parts of the world, such as Columbia, the temperatures, light, and other conditions are absolutely perfect for growing certain cut flowers.

About 90 percent of the cut roses produced in the United States are hybrid tea roses. The hybrid tea rose is a large, usually solitary flower. The sweetheart (floribunda) rose is a smaller rose that grows as a spray and is often disbudded. Propagation is done by grafting. Roses like an organic medium, high fertility, warm temperatures, and high light intensity. Pruning of roses every year or at least every other year helps maintain production levels. It takes from six to eight weeks to produce a new crop of roses after a harvest.

Gerbera daisies are bright flowers. Propagation is by tissue culture. Proper planting depth and well-drained soil are essential for reducing soil borne disease. Flowers are produced 8 to 12 weeks after plants are planted in beds.

Carnations rank as the third most popular cut flower in the world. Three types of cut carnations are grown: standard carnations, mini carnations, and single flowering types. Columbia has perfect conditions for year-round carnation production. Carnations prefer 50 to 55°F nights and 60 to 65°F days and high light intensity. The biggest problem with harvested flowers is the splitting of the calyx caused by fluctuations of temperatures.

Gladioli are long-lasting cut flowers. Much of the commercial production of gladioli takes place in Florida and California. Gladioli are produced outside in fields from corms. Harvested gladioli are kept in an upright position to reduce bending caused by gravitropism. Thrips can be especially troublesome and cause damage to both foliage and flowers.

Snapdragons are propagated by seed. The hybrids used have been developed for specific growing periods of the year based on temperature and light intensity. It is necessary to keep the flowers upright after harvesting, during storage, and while shipping to maintain straight stems. Snapdragons have few serious pest or disease problems.

Alstroemeria has gained popularity in the United States since the 1970s. It is ever blooming, produces high yields, and it is easy to grow under cool temperatures. Propagation of alstroemeria is done by division of its white fleshy rhizomes or by tissue culture. Alstroemerias are very heavy feeders. They also require large amounts of water and high light intensity for maximum floral production.

Freesia has brightly colored lily-like flowers that are very fragrant. Freesias are produced from corms. Freesias are planted in September in northern regions and as late as December in southern regions of the United States. Perlite must not be used in the medium, as freesias are sensitive to fluorides. The first flowers appear about eight weeks following the onset of cool temperatures.

QUESTIONS

Answer the following questions using complete sentences and correct spelling.

1. What has caused the decreased percentage of cut flowers produced by United States growers?
2. Which states are the greatest cut flower producers?
3. Why has cut flower production moved to other countries?
4. How do hybrid tea and sweetheart roses differ?
5. Why are roses pruned?
6. How do standard, mini, and single flowering carnations differ?
7. Why is Columbia a major producer of carnations?
8. Why are gladioli and snapdragons kept upright after harvest?
9. What are three crops discussed in this chapter that have specialized underground stems?
10. Which cut flower crop is grown primarily from seed?

EVALUATING

Match the term with the correct definition. Write the letter of the term on the line provided.

a. alstroemeria
b. freesia
c. gladiolus
d. hook cut
e. hybrid tea rose
f. mini carnation
g. single flowering carnation
h. splitting of the calyx
i. standard carnation
j. sweetheart (floribunda) rose

_____1. Bright, lily-like flowers borne on a horizontal spike that are very fragrant.

_____2. A smaller rose that grows as a spray and is often disbudded.

_____3. Large, solitary carnations that have been disbudded.

_____4. Ever-blooming flower borne on a cyme, produces high yields, and it is easy to grow under cool temperatures.

_____5. A carnation that is 1/2 to 1/3 the size of a standard carnation and is produced with two to four flowers in a spray.

_____6. The removal of a portion of the old wood with a flower when harvesting.

_____7. A solitary carnation flower produced without the need of disbudding.

_____8. A condition caused by fluctuations of temperatures resulting in a misshapen flower.

_____9. A large, usually solitary, rose.

_____10. Long-lasting, large, showy floral spikes from plants with corms.

EXPLORING

1. Visit a local florist or retailer selling fresh cut flowers. Identify the flowers being sold. Are the same flowers available in all the stores? Is there a greater diversity of flowers than you imagined? Less?

2. Arrange a trip to a wholesale florist. Make an appointment with the owner or manager so you can ask questions concerning the handling and care of fresh cut flowers. Determine where the flowers were grown. Ask how they were purchased? Find out how they were shipped and where they will be going.

3. Visit a local operation that produces cut flowers. On your fact-finding tour, learn about how the crop is produced and handled. Inquire as to whether the crops are sold locally or shipped elsewhere.

CHAPTER 13

Growing Foliage Plants

OBJECTIVES

This chapter is about growing foliage plants. It has the following objectives:

1. Recognize the scope of the foliage plant industry.
2. Identify major foliage plants.
3. Relate practices used in commercial foliage plant production.
4. Describe cultural practices in growing foliage plants.
5. Explain techniques for caring for foliage plants indoors.

TERMS

foliage plants

interior plantscaping

13–1. Foliage plants bring freshness and beauty indoors. (Courtesy, Jim Frost, Urban Flora, Inc.)

HERE is a challenge. See if you can go through a day without seeing a foliage plant. It is not easy. Most people cannot help but come in contact with foliage plants every day. They are grown in the home. People keep foliage plants on their desks at work. The interiors of malls are decorated with foliage plants. Foliage plants are everywhere!

Why are foliage plants used so extensively? One reason is that foliage plants bring a freshness and natural beauty to interior spaces. They filter pollutants and dust from the air. They produce oxygen via photosynthesis. They have the quality of being able to soften harsh features in buildings. They grow throughout the year. Most important, they are pleasant to look at.

FOLIAGE PLANT PRODUCTION

Foliage plants consist of tropical and subtropical plants selected for their ability to be grown indoors. They are commonly referred to as houseplants because of their wide use in residential homes. The color and interest provided by their leaves make them attractive to people. Some foliage plants produce interesting or colorful flowers that add to their value.

13–2. Some foliage plants have colorful leaves, like these crotons.

Foliage plants are grown for the purpose of interior plantscaping. ***Interior plantscaping*** involves the use of plant materials to improve the appearance of the indoor environment. Interior plantscaping usually implies a complex design involving many plants. However, use of a single plant or a single type of plant fits the definition.

13–3. Interior plantscaping involves the use of plant materials to improve the appearance of the indoor environment. (Courtesy, Jim Frost, Urban Flora, Inc.)

13–4. The John Deere Headquarters in Moline, Illinois, uses foliage plants extensively in their interior plantscape.

THE VALUE OF FOLIAGE PLANTS

Foliage plants are an important part of the floriculture industry. According to the USDA 2001 Summary of Floriculture Crops, foliage plant production ranks third behind bedding plants and potted flowering crops. The wholesale value of foliage plants was calculated to be $495 million in 2001. Furthermore, foliage plant production is expected to grow in future years.

Most foliage plants are produced in warm states. Florida produces more foliage plants than all the other states combined. California is a distant second, followed by Texas and Hawaii. The warm climates and qualities of light found in these states are ideal for growing

13–5. Foliage plant production ranks third behind bedding plant and potted flowering crop production in the United States.

13–6. Thousands of Spathiphyllum are being grown in this Florida operation.

13–7. Three types of Dracaenas are shown in the foreground of this photograph.

13–8. Pothos ranks second in production.

foliage plants. Heating is often not necessary, so production costs are less than they would be in a colder climate.

Twelve types of foliage plants dominate sales in the United States. Together, they make up 60 percent of the total market. The 12 include *Dracaena*, pothos, *Ficus*, dumbcane, palms, Chinese evergreen, peace lily, English ivy, *Philodendron*, *Schefflera* and *Brassaia*, nephthytis, and ferns.

13–9. A weeping fig, a rubber tree, and a variegated weeping fig represent three plants in the genus Ficus grown as foliage plants.

13–10. Two of the many cultivars of dumbcane that are available.

GENERAL INFORMATION

The production of foliage plant crops involves a number of steps. Production begins with the propagation of plant material. Then, plants are grown quickly under optimum conditions and prepared for sale. Once sold, foliage plants are packaged and shipped throughout the country to the retail florists, garden centers, interior plantscapers, and mass market distributors. These retailers have the responsibility of maintaining the health of the plants until they are sold to the public. Maintaining the health and appearance of the plant then rests with the customer.

13–11. Germination of seed is improved when flats are placed on tubing through which warm water circulates.

Propagation

Foliage plants are propagated in a variety of ways. The most common methods are stem cuttings and leaf cuttings. Other methods of propagation include seed, spores, separation, division, air layering, and tissue culture. The selection of the method of propagation depends largely on the species of plant being propagated.

13–12. A person transplanting palm seedlings to 4-inch pots.

Some foliage plants lend themselves well to propagation by seed. Palms are typically produced from seed. Sexual propagation is also common with Buddhist pine (Podocarpus macrophylla), Norfolk Island pine (Araucaria heterophylla), False-aralia (Dizygotheca elegantissima), Schefflera (Brassaia actinophylla).

13–13. This person is seen checking the development of Spathiphyllum tissue cultures.

Tissue culture has grown in importance as a method of propagating foliage plants. Thousands of identical plants can be pro-

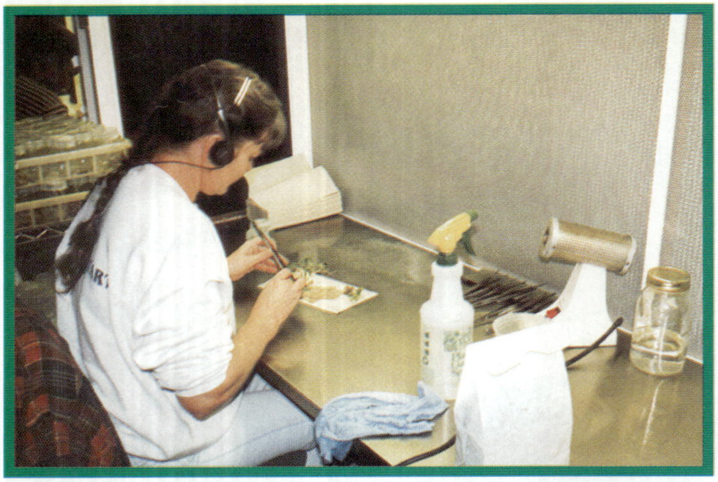

13–14. Under a laminar air flow hood, this person divides cultures before returning the explants to jars with multiplication media.

13–15. An explant. (Courtesy, *Greenhouse Product News*)

duced in a relatively short period. In addition, the plants produced are free of disease. Most ferns, including the Boston fern (*Nephrolepis exaltata* 'Bostoniensis'), are propagated through tissue culture. Peace Lily (*Spathiphyllum* hybrids), dumbcane (*Dieffenbachia* species), and Nephthytis (*Syngonium podophyllum*) are also easily propagated by tissue culture.

Mother-in-law's tongue (Sansevieria) is divided because varieties do not reproduce true to form if propagated by cuttings. This is a disadvantage because of the labor involved in division. Air-layering is another propagation method. It was once used widely with Schefflera and Fig (*Ficus* species). Its use has been in decline because of the intense labor required.

13–16. Small plants produced through tissue culture are moved to the greenhouse for production.

13–17. This worker sticks English ivy cuttings directly into the finish containers, thereby saving the labor involved in transplanting.

Mist systems have greatly improved propagation success. Cuttings root quickly under mist because plants are kept cooler and water stress is reduced. Many growers save labor by sticking cuttings directly into the finish containers.

Growing

As with other greenhouse crops, the health of foliage plants depends largely on cultural practices. Appropriate growing medium, light, water, nutrition, temperature, and air quality are key factors in production. Attention to pest and disease problems is also important.

Growing Medium

Growing media are prepared with the needs of the plants in mind. Certain plants, including some ferns, like a 100 percent organic medium. Succulents and cacti prefer a medium with 50 percent or less organic material and a greater amount of sand or mineral matter that provides good drainage. However, the vast majority of foliage plants do well in commercially prepared growing media. Most foliage plants like a slightly acid pH falling

13–18. Cacti and succulents prefer very well drained growing media. (Courtesy, *Greenhouse Product News*)

13–19. These workers use a commercially prepared growing medium for dish gardens.

between 5.5 and 6.5. Exceptions include the prayer plant (*Maranta*) and ferns. Both do best with a more acidic (4.5 to 5.5) medium.

Light

Foliage plants have been selected for their ability to withstand low light conditions. Most are found growing on forest and jungle floors. The light intensity for which they are adapted in nature is not too different from the light intensity they receive in an interior plantscape.

Light requirements of most foliage plants fall between 1,500 and 8,000 foot candles. Those levels are lower than what is found in a typical greenhouse. Therefore, most foliage

13–20. These crotons are being grown under shade cloth to reduce the intensity of the light. They are elevated to reduce ant problems.

plants are protected from the intense light of full sun during production. Exposure to full sun can cause damage to the leaf tissues and chlorophyll pigments. A symptom of excessive light is faded, or bleached, leaf colors.

Temperature

Since most foliage plants are tropical or subtropical in origin, they are accustomed to relatively warm and constant temperatures. The best temperatures for the production of most foliage plants are 65°F to 80°F nights and 75°F to 95°F days. Temperatures above 95 °F can damage crops. If high temperatures are a concern, shading can be used to help protect foliage plants. Temperatures below 65°F slow plant growth and delay plant development.

Water

To maintain optimum growth, water should be available to plants at all times. A number of environmental conditions influence the frequency of watering. Watering may be necessary daily or weekly. Ebb and flood, spaghetti tubing, and capillary mats work well with foliage plants. Irrigation methods that wet the leaves invite disease problems and spotting. As a result, overhead watering is often avoided.

Watering practices depend largely on the species being grown. Growers must understand the watering needs of the plants they grow. Some foliage plants like a medium that is kept moist and not allowed to dry. Some prefer medium that is allowed to approach drying before irrigation. Others do best when the medium is allowed to dry completely before watering. Whatever the plant prefers, the rule is to water heavily when it is time to water. Heavy watering leaches, or washes, soluble salts from the medium and insures that the medium is completely wetted.

13–21. This interior plantscaper determines when a plant is in need of water. (Courtesy, Jim Frost, Urban Flora, Inc.)

Improper watering is one of the major causes of root diseases of foliage plants. To avoid root rot diseases, start with the appropriate growing medium for the plant and the right size container for the plant. When watering, wait until the plant needs water before watering.

Most foliage plants like a humid environment. In fact, humidity of 50 percent or more is common where foliage plants grow in nature. A relatively high

13–22. Ferns are one of many foliage plants that like a humid environment.

humidity level in the greenhouse improves growth. Growth tends to slow if humidity levels drop below 25 percent.

Nutrition

Nutrition during production influences the rate of growth. It also affects the quality of foliage plants after they reach the consumers. One recommended fertilizer program consists of a constant feeding program. With every watering, 150 ppm nitrogen, 25 ppm phosphorus, and 100 ppm potassium are provided. An alternative is to use a slow-release fertilizer in the growing media. Foliage plants grown in soilless medium require fertilizers containing micronutrients. The reason is soilless mix lacks the necessary micronutrients.

Acclimatization

Acclimatization is the process in which a plant becomes accustomed to a new environment. The transition from a warm, humid, well-lit greenhouse to a cool, dry, and dim room in a building is stressful for most plants. A common reaction by a plant under stress from a new environment is to shed its leaves. Few plants demonstrate the shock of being moved as much as the weeping fig. A slight change in conditions causes it to shed most of its leaves.

13–23. Growers acclimatize foliage plants before shipping. (Courtesy, Cheryl Schroeder)

Studies have shown that the plants that acclimatize the quickest are those grown under low light conditions and that receive light fertilization throughout production. Knowing this, growers can help prepare plants for their ultimate destination.

First, the light intensity given the plants can be reduced for two to six months prior to shipping. The length of time depends on the species. By reducing the light intensity, the plants get accustomed to lower light levels expected in their new home. Their leaves grow thinner, broader, and produce more chlorophyll for the manufacture of food.

Over fertilization during production lessens the plant's ability to acclimatize to a new surrounding. Fertilization stimulates growth. Foliage plants acclimatize more quickly if growth is slowed. Therefore, to better prepare plants, growers reduce fertility levels to slow plant growth.

Table 13–1. Growing Guidelines for Selected Foliage Plants

	Light	Water	Humidity	Problems
Asparagus fern *Asparagus sprengeri*	Medium 2,500 to 4,500 FC	Let dry slightly	Medium	Spider mites
Baby's tears *Helxine soleirolii*	Medium 2,000 to 4,000 FC	Keep moist	High	
Begonias *Begonia* species	Medium 2,500 to 5,000 FC	Let dry slightly	Medium	Mites
Brassaia (Schefflera) *Brassaia actinophylla*	Medium 3,000 to 5,000 FC	Let dry slightly	Medium	Mites, Mealybugs, Scale, Thrips
Bromeliad *Neoregelia* species	Medium 2,500 to 4,000 FC	Let dry slightly	Medium	
Cacti Many species	High 4,000 to 8,000 FC	Let dry	Low	
Cast iron plant *Aspidistra elatior*	Low 1,500 to 3,000 FC	Let dry slightly	Medium	
Chinese evergreen *Aglaonema* species	Low 1,000 to 2,500 FC	Keep moist	Medium	Mealybugs, Scale
Croton *Codiaeum* species	High 3,000 to 8,000 FC	Let dry slightly	Medium	Mites, Mealybugs, Scale
Dracaena *Dracaena* species	Medium 2,000 to 4,000 FC	Let dry slightly	Medium	Mites, Mealybugs, Thrips
Dumbcane *Dieffenbachia* species	Low 1,500 to 2,500 FC	Let dry slightly	Medium	Mites, Mealybugs, Scale
English ivy *Hedera helix*	Low 1,500 to 2,500 FC	Let dry slightly	Medium	Mites, Scale
False aralia *Dizygotheca elegantissima*	Medium 2,000 to 4,000 FC	Let dry slightly	Medium	Mites, Mealybugs, Scale
Ferns Many species	Low 1,500 to 3,000 FC	Keep moist	High	Scale, Caterpillars, Mealybugs

(Continued)

Table 13–1 (Continued)

	Light	Water	Humidity	Problems
Figs *Ficus* species	High 4,000 to 8,000 FC	Let dry slightly	Medium	Mite, Mealybugs, Scale, White fly
Fittonia *Fittonia verschaffeltii*	Low 1,000 to 2,500 FC	Let dry slightly	High	Mealybugs
Grape ivy *Cissus rhombifolia*	Low 1,500 to 2,500 FC	Let dry slightly	Medium	Mites
Jade plant *Crassula argentea*	High 6,000 to 8,000 FC	Let dry	Low	Mealybugs, Scale
Mother-in-law's tongue *Sansevieria* species	Adaptable 1,500 to 6,000 FC	Let dry	Low	
Nephthytis *Syngonium podophyllum*	Low 1,500 to 3,000 FC	Let dry slightly	Medium	Mites, Mealybugs
Norfolk Island pine *Araucaria heterophylla*	High 4,000 to 8,000 FC	Let dry slightly	Medium	Mites
Palms Various genus	Low-medium 3,000 to 6,000 FC	Keep moist	Medium	Mites, Mealybugs, Scale
Peace lily *Spathiphyllum* species	Low 1,500 to 2,500 FC	Let dry slightly	Medium	Mealybugs, Scale
Peperomia *Peperomia* species	Low 1,500 to 3,000 FC	Let dry slightly	Medium	
Philodendron *Philodendron* species	Low 1,500 to 3,500 FC	Let dry slightly	Medium	Mites, Mealybugs, Scale, Thrips
Pilea Many species	Low 1,500 to 2,500 FC	Keep moist	Medium	
Pothos *Epipremnum* species	Low 1,500 to 3,000 FC	Let dry slightly	Medium	Mealybugs, Thrips
Prayer plant *Maranta* species	Low 1,500 to 3,500 FC	Keep moist	High	Mites, Mealybugs
Purple passion vine *Gynura sarmentosa*	Medium 3,000 to 5,000 FC	Let dry slightly	Medium	
Schefflera *Schefflera arbicola*	Low 1,500 to 3,000 FC	Let dry slightly	Medium	Mites, Mealybugs, Scale, Thrips
Spider plant *Chlorophytum comosum*	Low 1,000 to 2,500 FC	Let dry slightly	Low	
Strawberry begonia *Saxifraga sarmentosa*	Medium 3,000 to 5,000 FC	Let dry slightly	Medium	

(Continued)

Table 13–1 (Continued)

	Light	Water	Humidity	Problems
Swedish Ivy *Plectranthus australis*	Medium 3,000 to 5,000 FC	Keep moist	Medium	Mealybugs
Ti plant *Cordyline terminalis*	Low 1,500 to 3,500 FC	Keep Moist	Medium	
Wandering Jew *Zebrina pendula*	Medium 2,500 to 4,000 FC	Let dry slightly	Medium	
Wax plant *Hoya carnosa*	Low 1,500 to 3,000 FC	Let dry	Medium	Mealybugs
Yucca *Yucca elephantipes*	Medium 3,000 to 5,000 FC	Let dry	Low	
Zebra plant *Aphelandra squarrosa*	Low 1,000 to 1,500 FC	Keep moist	Medium	Mites

Pests and Diseases

Common pests of foliage plants include aphids, mites, caterpillars, fungus gnats, mealybugs, scales and thrips. Pest problems can be reduced with the use of pest-free stock plants. Sanitation and cultural practices also help control pest problems.

Disease problems of foliage plants are most often associated with watering practices. Over watering, usually caused by watering too frequently, often leads to root rot. Fungal

13–24. Foliage plants grown in Florida are packaged and shipped north.

13–25. Quality foliage plants are free of pests and diseases. (Courtesy, Society of American Florists)

and bacterial diseases can attack foliage. They become more of a problem when the leaves are wet.

Fluoride in the water is damaging to some plants. Dracaena, palms, prayer plant, spider plant, and bromeliads are susceptible. Fluoride damage is usually in the form of leaf tip burn or browning. Avoid growing media with perlite and the use of super phosphate fertilizer.

13–26. The sale of these cacti was promoted at Christmas.

13–27. This person disinfects previously used flats before they are used for the propagation of foliage plants.

GROWING FOLIAGE PLANTS INDOORS

Foliage plants can be grown successfully indoors by following a few simple guidelines.

- Select foliage plants based on their ability to withstand the growing conditions.
- Provide optimum lighting for the foliage plants. Place plants requiring medium to high light near windows. North facing windows have the lowest light intensity. South facing windows provide the brightest light in the winter when the sun is low in the horizon. The further a plant is placed away from a window the lower the light intensity will be.
- Maintain temperatures above 55°F.
- Employ the use of a humidifier in the winter months. Humidity drops significantly in houses with forced-air heat.
- Select containers that are in proportion to the size of the plants to be grown. Plants grown in containers too small, dry quickly. Those in containers too large are subject to root rot. The volume of growing medium in large pots is slow to dry leading to root rot diseases. For root health, select containers with holes that allow drainage of excess water.
- Repot plants when the roots completely fill the soil ball and if additional growth of the plant is desired. Repotting to a larger container stimulates growth. When repotting, plant the plants in the next size container. Use sterile growing medium. Do not reuse growing medium as it can contain disease organisms and high levels of soluble salts.
- Water when the plant needs it, not on a fixed schedule, such as every Saturday. Water thoroughly, and do not allow the plants to stand in water.
- Fertilize sparingly to maintain growth. Use of a complete fertilizer every four to six weeks is adequate.

13–28. Select foliage plants on their ability to withstand the growing conditions. (Courtesy, Society of American Florists)

REVIEWING

MAIN IDEAS

Foliage plants, commonly referred to as houseplants, consist of tropical and subtropical plants selected for their ability to be grown indoors. Foliage plants are grown for the purpose of interior plantscaping. Their production in the United States places them third behind bedding plants and potted flowering crops. The state with the greatest production is Florida.

Twelve types of foliage plants make up the majority of the foliage plants grown. They include *Dracaena*, pothos, *Ficus*, dumbcane, palms, Chinese evergreen, peace lily, English ivy, *Philodendron*, *Schefflera* and *Brassaia*, nephthytis, and ferns. Foliage plants are propagated sexually and asexually. The most common methods are stem cuttings and leaf cuttings. Other methods of propagation include seed, spores, separation, division, air layering, and tissue culture.

Cultural practices vary with the type of plant being grown. However, most foliage plants like an organic growing medium with a slightly acid pH falling between 5.5 and 6.5. Light requirements of most foliage plants fall between 1,500 and 8,000 foot candles. The best temperatures for the production of most foliage plants are 65 to 80°F nights and 75 to 95°F days. Watering practices depend largely on the species being grown. A recommended fertilizer program consists of a constant feeding program of 150 ppm nitrogen, 25 ppm phosphorus, and 100 ppm potassium.

It is important that foliage plants be acclimatized. Acclimatization is the process in which a plant becomes accustomed to a new environment. Plants properly acclimatized tolerate new growing environments much better than those that are not acclimatized.

QUESTIONS

Answer the following questions using complete sentences and correct spelling.

1. What is interior plantscaping?
2. Why does much of the foliage plant production take place in southern states?
3. How has tissue culture grown in importance for propagating foliage plants?
4. What foliage plants are propagated in large numbers by tissue culture?
5. What are the advantages of mist systems in production?
6. Why are watering methods that wet leaves avoided?
7. How do leaves respond to lower light levels?
8. Why is it important to limit fertilization of foliage plants?
9. How does fluoride affect foliage plants?
10. What is the recommended watering practice for foliage plants in an interior situation?

EVALUATING

Complete the following lists on a separate sheet of paper.

1. List 12 foliage plants that have few if any pest and disease problems.
2. List nine foliage plants that like the growing medium to be kept moist.
3. List five plants that require high levels of light.
4. List five foliage plants that tolerate low light levels.
5. List five foliage plants that prefer the growing medium to dry out between waterings.
6. List 10 foliage plants that can become infested with mealy bugs.
7. List eight foliage plants that do well under medium light and like having the growing medium dry slightly between waterings.
8. List four plants that require humid conditions.
9. List five plants that tolerate low humidity.
10. List seven practices a homeowner can follow to maintain their foliage plants.

EXPLORING

1. Visit a large conservatory. As you walk around, observe how foliage plants might grow differently in that setting versus the home environment.

2. The best way to learn how to take care of foliage plants is to care for foliage plants. Obtain foliage plants of different types and care for them in your home. Note how some plants might perform better that others. Consider the growing conditions in your home and the types of plants you have. Then determine why they perform as they do.

CHAPTER 14

Growing Bedding Plants

OBJECTIVES

This chapter is about growing bedding plants. It has the following objectives:

1. Identify major bedding plants.
2. Describe cultural practices applied to the production of bedding plants.
3. Explain how to start bedding plants from seed.
4. Explain how plugs are used in bedding plant production.
5. Prepare a growing schedule for select bedding plants.
6. Discuss perennial production as part of a greenhouse operation.

TERMS

annuals
bedding plants
break
cell packs
DIF
fertigation
finished
germination rate
hard basket
leaching
medium
perennials
plugs
shelf life
soft basket
sow
toning
transplanted
viable

14–1. Bedding plants add color to the landscape. (Courtesy, *Greenhouse Product News*)

PEOPLE love plants that provide color around the home. Cheerful colors stimulate and excite emotions. Flowers and foliage are especially welcome in states where inhabitants have experienced a long, cold winter. The arrival of spring draws people outside where they can reacquaint themselves with nature. Naturally, people want the best display of flowers for the longest period of time. Trees and shrubs may have showy displays of flowers, but they seldom last more than a few weeks. Bedding plants, on the other hand, provide color from the time they are planted until a killing frost.

Planting bedding plants is one of the most common leisure activities in the country. It is easy! The gardener is also nearly always rewarded with beautiful displays of flowers and foliage.

WHAT ARE BEDDING PLANTS?

Bedding plants include herbaceous annual plants used for ornamental display or vegetable production. Frequently, people will refer to bedding plants as ***annuals***. Annuals are herbaceous plants that grow from seed to flower in one season before dying. Bedding plants are selected for their colorful flowers or interesting foliage. A few examples of the bedding plants produced for their flowers include geraniums, petunias, impatiens, and marigolds. Coleus is the most common bedding plant grown for its foliage or colorful leaves. Typically, the vegetable plants included as bedding plants are those started in containers and transplanted to the garden. Some of these include tomatoes, peppers, broccoli, cabbage, and Brussels sprouts.

Table 14–1. Value of Bedding Plants Produced in the United States by Crop in 2001, USDA (in $1,000s)

Crop	Flats	Pots	Hanging Baskets
Begonias	60,374	21,362	18,847
Geraniums, cuttings	12,240	111,870	28,249
Geraniums, seed	8,135	37,986	3,748
New Guinea Impatiens	5,723	38,530	29,540
Impatiens	117,476	26,598	18,036
Marigolds	51,152	5,740	650
Pansy	103,179	20,855	2,725
Petunia	96,408	22,124	18,524
Other	340,489	290,217	96,089
Vegetables	72,021	28,697	—

TRENDS WITH BEDDING PLANTS

The market for bedding plants is bright. Sales are up and the demand is rising. According to the USDA, the wholesale value of bedding and garden plants totaled $2.18 billion in 2001. This represented 49 percent of the wholesale value of all floriculture crops grown in the United States. California, Michigan, Texas, Ohio, and Florida accounted for 42 percent of the total wholesale value of bedding plants. Overall, potted geraniums (from both cuttings and seed) sold for the highest value to growers. Impatiens flats provided the second largest amount. Clearly, the production and sales of bedding plants have very important roles in the floriculture industry.

Many factors have spurred the increased sales in the bedding plant industry. One factor is marketing. At one time, consumers were limited to garden centers or greenhouses from which to buy their bedding plants. Today, bedding plants are sold nearly everywhere. Large retail

14–2. Bedding plant sales are up.

14–3. Large retail chains sell bedding plants.

chains, such as Wal-Mart, Home Depot, and Kmart, have expanded the market. Grocery stores around the country serve as outlets for bedding plant sales. Yet, garden centers remain the top outlets for quality bedding plants.

Research and development have strengthened the industry. Many new varieties resulting from plant breeding programs give consumers higher quality plants and more types from which to choose. Technology and automation have increased the number of plants producers can grow. Biotechnology and genetic engineering are expected to greatly influence the industry, too. Genetically engineered crops will be more resistant to disease and insect pests, will improve floral displays, and will reduce fertilizer requirements because plants would fix their own nitrogen.

The consumer has also played a key role in determining which crops are grown. The United States has experienced a long period of economic prosperity. People are willing to use more disposable income on garden plants. Hence, the demand for bedding plants is high. Consumers have also become more knowledgeable and are willing to try new varieties.

BEDDING PLANT SCHEDULING, SEED, AND GERMINATION

The majority of bedding plants are grown from seed or propagated sexually. One common exception is the zonal geranium, which is usually propagated asexually from stem cuttings. Seeds come in a wide variety of shapes and sizes. The fibrous begonia seed

14–4. Most bedding plants, including impatiens, petunias, and pansies are grown from seed. (Courtesy, Ball Horticultural Co.)

is so small it looks like dust, and an ounce of the seed would include an incredible 2,500,000 seeds. Others, like the salvia, are larger, 7,500 seeds to the ounce, and are easier to handle.

Growers **sow**, or plant, bedding plant seed in late winter or early spring. There are terrific benefits to starting plants early. The rate of seed germination and seedling survival is much greater in the controlled environment of a germination chamber or greenhouse than it is outside. Bedding plants started in a controlled environment receive optimal temperatures, water, and nutrients important for healthy growth and development. The healthy transplants therefore have a head start. Also, by planting a healthy and vigorously growing plant in the garden, the gardener can enjoy a longer season of flowering pleasure.

14–5. Temperature and humidity can be controlled in a germination chamber.

SCHEDULING

Production of bedding plant crops really starts with a schedule. The growers must ask when they want the plants for sale based on the consumer. Plants started too soon may be overgrown and less valuable during the prime period of sales. Plants started too late may be too small to attract customers.

Table 14–2. Finishing Times for Bedding Plants

	Number of Weeks to Finish From 390 Plugs			Number of Weeks to Finish From Seed		
	Flats		Pots	Flats		Pots
	48	36	4 inch	48	36	4 inch
Ageratum	5–6	7	8	10	11	13
Alyssum	5–6	7	8	8	9	11
Begonia	6	7	8	15	16	18
Browallia	6	7	7	15	16	20
Celosia	5	6	7	8	9	13
Cleome	3	3	4	9	10	12
Coleus	4–5	5	5	9	10	12
Dahlia	9	10	11	12	13	14
Dianthus	5–6	7	8	10	11	15
Dusty miller	5–6	6	7	11	12	15
Gazania	7	8	9	12	13	15
Geranium	10	10	11	13	14	15
Impatiens	5	5	6	10	11	13
Lisianthus	10	10	11	18	19	20
Lobelia	4–6	7	8	11	12	12
Marigold (French)	4–5	5	6	8	9	10
Marigold (African)	5–6	6	7	11	12	13
Pansy	5–6	7	8	14	15	16
Petunia	5–6	6	6	12	13	15
Portulaca	5	6	7	12	13	15
Salvia	5	6	7	9	11	13
Snapdragon	5–6	6	7	15	16	17
Verbena	6	7	7	12	13	15
Vinca	5–7	8	8	14	15	16
Viola	6	7	8	12	13	14
Zinnia	2–3	3	4	9	10	12

14–6. A great deal of planning goes into scheduling sowing dates to meet customer needs.

Growers determine a date at which they want the plants to have their peak appearance and health for sale. They then count backward the number of weeks it takes to produce the crop. For instance, if growers know they want to have impatiens ready for a May 14 sale and it takes 10 weeks to grow the crop from seed, they would sow the seed 10 weeks before May 14. When sales are over a period of time, impatiens might be started every week for a period of four to six weeks to maintain a steady supply.

14–7. The pansies on this bench are ready to be shipped for sale. Note the use of the growing space in the greenhouse, from the plants on the floor to the hanging baskets on the ceiling.

SEEDS

Bedding plant seeds are expensive. Large greenhouse operations might purchase $100,000 of seed for production. Because of the seed cost, it is common for growers to lock their seeds up at night. The value of the seed also makes it necessary to provide proper storage. Proper storage extends the life of the seed. Growers store the seeds in cool dry locations to keep the seeds viable. Storage with a temperature of 40°F and a relative humidity between 20 and 40 percent is acceptable. Small growers often place their seed in sealed containers in a frost-free refrigerator.

14–8. Seeds can be kept sealed in their packets in a cooler.

Proper storage is one factor that contributes to a high rate of seed germination and plant survival up to sale. Germination is the sprouting of seeds or the beginning of plant growth. Successful germination begins with good seed. Good seed is viable and has vigor. **Viable** is a term used for living seeds capable of germinating under optimum conditions. Vigor describes the seed's ability to germinate under less than ideal conditions.

Fortunately, research has improved seed quality. Most seeds packaged for the current year have a **germination rate** of 90 to 95 percent when given optimum germination conditions. The germination rate is the percentage of seeds capable of germinating. Germination rates drop as the seeds get older. Seeds left over from previous years have noticeably lower germination rates than fresh seed as some seeds die in storage.

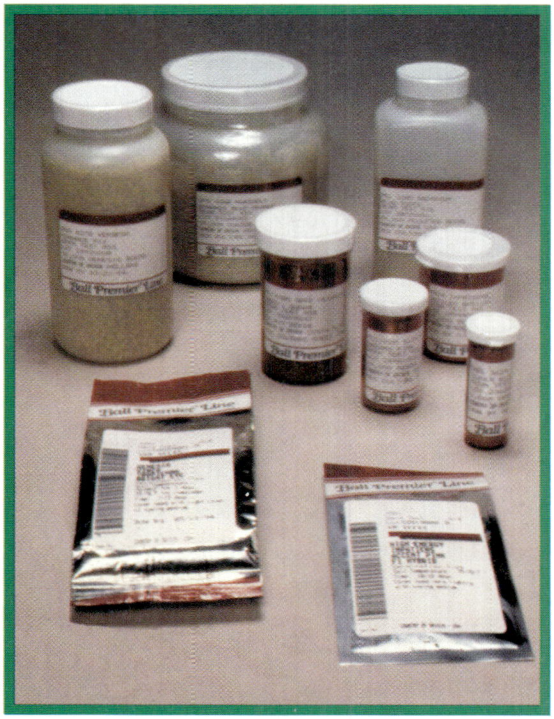

14–9. Germination rates are printed on the seed package. (Courtesy, Ball Horticultural Co.)

GERMINATION REQUIREMENTS

Always check for special germination requirements with each type of seed. Germination requirements include light, temperature, nutrients, moisture, and medium. Each germination requirement can be controlled by the grower for each bedding plant.

Table 14–3. Germination Guidelines for Bedding Plants

Common name	Seeds/ounce	Optimum temp. for germination	Light or dark exposure	Days for uniform germination
Ageratum	200,000	78–82	light	8–10
Alyssum	90,000	78–82	light	5
Begonia	2,500,000	75–80	light	5–10
Browallia	125,000	72	light	7–15
Celosia	35,000	75	dark	8–10
Cleome	14,000	80 days, 70 nights	dark	10–12
Coleus	100,000	70–75	light	10–14
Dahlia	4,500	60–65	no effect	5–10
Dianthus	25,000	70–75	no effect	7
Dusty miller	7,000	72–75	light	10–15
Gazania	12,000	70	dark	10–12
Geranium	6,000	70–75	no effect	10
Impatiens	46,000	70–75	light	4–5
Lisianthus	624,000	70–75	light	10–20
Lobelia	1,000,000	75–80, then drops to 70	no effect	20
Marigold (French)	9,000	72–75	no effect	7
Marigold (African)	9,000	72–75	no effect	7
Pansy	20,000	65–75	dark	7–10
Petunia	245,000	75–78	light	10–12
Portulaca	280,000	75–80	dark	10
Salvia	7,500	75–78	light	12–15
Snapdragon	180,000	70–75	light	7–14
Verbena	10,000	65	dark	20
Vinca	21,000	78–80, after three days drops to 75–78	dark	7–15
Viola	30,000 to 40,000	65–70	dark	7–14
Zinnia	2,500	70–72	no effect	3–7

Exposure to light influences seed germination. Some seeds must have darkness to germinate. Bedding plant seeds requiring darkness include celosia, cleome, gazania, periwinkle (Vinca), portulaca, pansy, verbena, and viola. It may therefore be necessary to see that the seeds are well covered. Other seeds require exposure to light. Seeds that need light to germinate include snapdragon, begonia, browallia, impatiens, sweet alyssum, petunia, and salvia. Still, others do not seem to be affected by exposure to light.

Proper temperature is important for successful germination. Most seeds germinate quickly with a medium temperature of 75°F. It is important to note that temperature refers to the soil temperature and not the air temperature. Check the soil temperatures with a soil thermometer. Soil temperatures that are too cool remain wet for a longer period of time. Wet, cool soil conditions reduce oxygen exchange with the seed and slow germination.

Seeds are started in a special **medium**, a soil or soilless material. The media used for germination purposes are easily purchased as prepackaged mixtures. The mix is fine-textured, uniform, and often consists of a peat moss and vermiculite. It is also free of disease organisms and weed seeds. Medium plays a minor role in germinating seeds. The key is the moisture level within the medium. Most seeds germinate well in a medium with a pH between 5.5 and 5.8. Geraniums, salvia, and marigold like a pH from 6.2 to 6.5.

14–10. These seedlings are two weeks old. (Courtesy, Ball Horticultural Co.)

The most important factor in seed germination is the moisture level in the medium. A rule of thumb is that once the medium and seeds have been moistened they should be kept moist until sprouting. If the medium is allowed to dry out during the germination process, the seeds die. It is best to wet the medium before sowing. After the seeds have been sown, the flats can be covered with a clear plastic cover or sheets of poly to reduce evaporation. The coverings are removed once the seeds start to sprout. Some growers mist their flats regularly to maintain proper moisture levels.

Different seeds prefer slightly different moisture levels. Seeds that prefer a wet medium or one that glistens and is wet to the touch are coleus, begonia, and alyssum. Moist medium that is wet, but not saturated, is prescribed for impatiens, petunia, geranium, salvia, vinca, and pansy. A dryer medium is best for verbena, zinnia, and most perennials.

In general, fertilizers are not used in the germination process. Fertilizers are soluble salts in the medium. Soluble salts inhibit the germination of most bedding plant seeds. Two exceptions are petunia and snapdragon. Low fertility levels actually enhance seed germination for those two crops.

HAND SOWING SEEDS

The depth at which the seeds are planted varies with the size of the seed and their light requirements. Larger seeds have more stored energy in their cotyledons enabling them to push up through the medium as they germinate. Tiny seeds have limited stored energy, and therefore lack the ability to push up through the soil if planted too deep. A rule of thumb is to plant the seeds at a depth no greater than three times the diameter of the seed. Hand sown seeds are placed evenly in rows rather than being broadcast over the entire flat. This practice protects the loss of all the seedlings by limiting the spread of disease if introduced.

After the seedlings develop their first true leaves, they are ***transplanted*** or moved to pots or cell packs. ***Cell packs*** are molded plastic containers divided into two, three, four, or six separate growing compartments. It is in the pots or packs that they will be ***finished*** or grown to saleable size. During the transplanting process, the seedlings must be carefully

14–11. Prepackaged medium is being placed in a machine that automatically fills flats.

14–12. Seeds sown by hand are placed in rows.

divided to avoid damaging the plant. It is recommended that the technician only handle the seedlings by their leaves since the stems are very easily bruised.

14–13. Seedlings should be held by their leaves during the transplanting process.

THE PLUG REVOLUTION

Propagation of bedding plants has changed dramatically with the industry move toward automation and the increased size of greenhouse operations. Small operations may still rely on hand seeding, but many small operations have moved away from seeding all together. Instead of propagating their own plants, they purchase plugs from large companies. The trays of plugs are easily custom grown for the medium and small greenhouse growers. Today, roughly two-thirds of all the seeded crops begin as plugs. It is safe to say that plugs have revolutionized bedding plant production worldwide.

Plugs are small plants grown in a small amount of medium in divided trays. The cells in the trays determine the size and number of plugs to be produced. The sizes of the plugs are 5/8 of an inch or larger. Plugs are very convenient as they can be popped out of the

14–14. This greenhouse is full of hundreds of plug trays sown in a day by automation.

14–15. Plugs are produced in plastic trays.

tray for transplanting to cell packs or pots. Plugs also allow for automated transplanting. The number of plugs grown in a tray varies. Tray designs may allow the grower to grow 800, 512, 390, 384, 288, 162, 144, or 70 plants per tray.

The larger producers of bedding plants use automatic seeders to produce plugs. The automated seeders greatly increase the speed at which plants can be seeded and the number of flats that can be produced. Automatic seeders place a seed in each cell of a plastic plug tray that has been filled with a germination mix or medium. Some growers use more than one seed per cell, especially

14–16. This automatic seeder places one seed (yellow) into each cell.

when germination rates are less than 85 percent. Petunias and begonias may be double-seeded to get a fuller tray. Five to six seeds per cell is common for portulaca or alyssum.

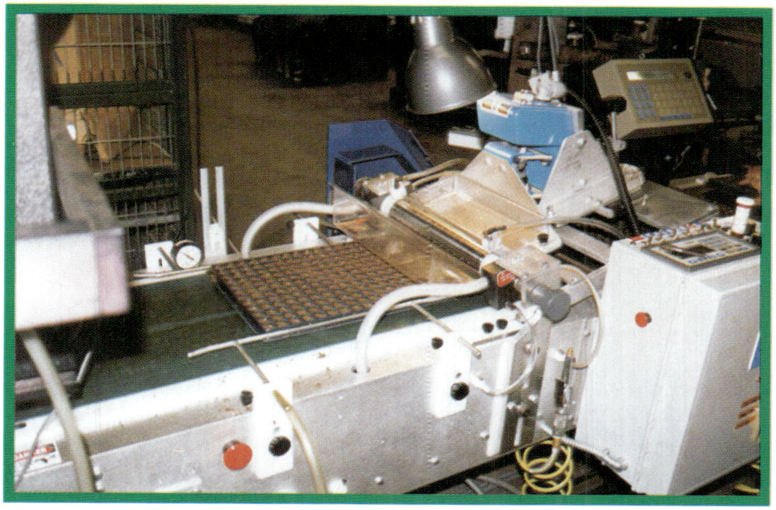

14–17. Seeders are computerized.

The materials used in the medium are extremely important. The most common material used is peat moss. It has an acidic pH, usually 4 to 5. Peat moss also absorbs and holds water. Vermiculite is another common plug medium ingredient. It retains water, it has good nutrient exchange, and it provides calcium and magnesium. Perlite is sometimes used to provide aeration. Calcined clay is used for aeration and for its nutrient exchange capacity. In some cases, soil is used in the medium. Lime is added to bring the pH levels up. The optimum pH for most bedding plants is 5.5 to 5.8. Geraniums, marigolds, and salvia prefer a higher pH of 6.2 to 6.5.

Different growers use different mixes. One example of a medium used consists of 80 percent peat, 13 percent vermiculite, 7 percent soil, with lime and fertilizer added. The mix has excellent moisture-holding and aeration capabilities. After the seed is sown, the medium is covered lightly with vermiculite and watered. The key to successful and uniform germination is a medium that provides proper moisture levels and is well aerated. Each tray is labeled with the plant variety to

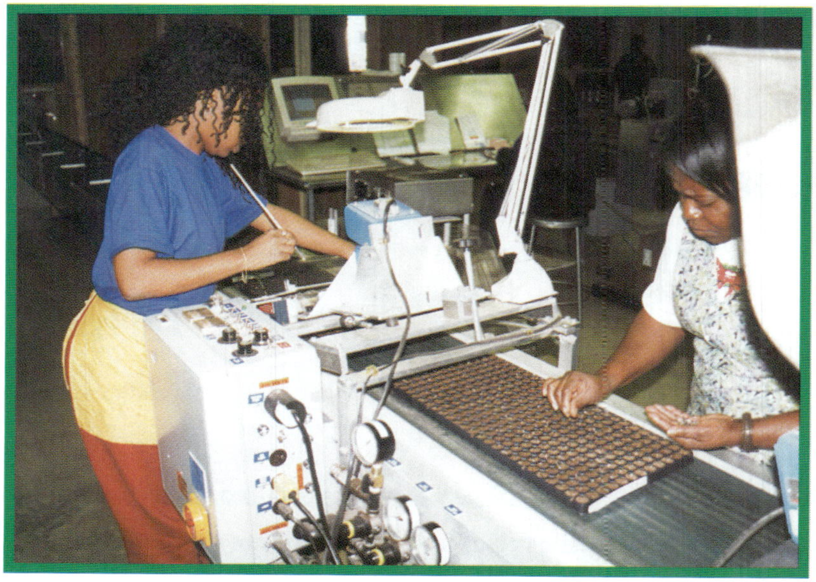

14–18. The person on the right plants seeds in cells the automatic seeder missed.

14–19. A light covering of vermiculite has been placed over these trays with marigold seeds.

14–20. Plug trays move on a conveyer belt through the seeder and then through this mechanism that waters the tray.

avoid confusion. Some growers then place the plug trays in specially designed germination chambers that provide the optimum temperatures for seed germination.

The goal of the grower is to produce plugs that are uniform, have short internodes, and have a high number of dark green leaves. To accomplish this, the plants move through four growing stages. The type of plant determines how long each stage lasts. Stage one involves germination of the seed. In stage two, the plants are kept warm and given low levels of fertilizers. Stage three involves higher levels of fertilizer to promote growth and development. In stage four, plants are prepared for shipping. To slow growth prior to shipping, the plants' fertilizer levels are reduced and they may be treated with a growth regulator. The entire process of producing quality plugs takes 6 to 10 weeks.

14–21. This person replaces plugs in which seed did not germinate with live plugs.

14–22. This photograph shows an automatic watering system involving a boom that moves across the entire length of the bench.

14–23. This photograph shows tomatoes wilting along the edges of the tray.

14–24. This person spot waters edges of trays that have dried quickly.

14–25. These plugs were treated with a growth retardant and a warning sign was posted.

14–26. Plugs are moved to greenhouses with retractable roofs prior to shipping so they can be hardened off by exposure to lower humidity, higher light intensity, and lower temperatures.

14–27. These Florida-grown plug trays are packaged for shipping to fill orders around the country.

GROWING BEDDING PLANTS

At the time of sale, high quality bedding plants are compact, with numerous bottom breaks and with the flower buds just ready to open. The term ***break*** refers to the development of lateral branches that contribute to a bushy appearance. Some vendors require bedding plants to be in flower. The primary factors in producing quality bedding plants are temperature, soil moisture, light, and fertilization. Growth regulators may also be used by growers to produce quality bedding plants.

14–28. High quality bedding plants are compact plants, with numerous bottom breaks and flower buds just ready to open at sale. (Courtesy, *Greenhouse Product News*)

TRANSPLANTING

Plugs are transplanted to cell packs or pots for growth until sale. One of the trends in the industry has been the move to larger finishing containers. Four-inch pots have become common. Most growers that use cell packs choose ones that hold no more than 48 plants in a standard flat. Higher quality plants can be produced when the larger containers are used. In addition, the plants require less frequent watering and have a longer shelf life. ***Shelf life*** is the period the plant maintains its health while on display for sale.

14–29. Many bedding plants are sold in plastic cell packs.

14–30. Labor is reduced by machines that fill pots with growing medium.

14–31. This machine adds osmocote, a slow-release fertilizer, to the growing medium before filling flats.

14–32. Pots filled with soil and plug trays move on conveyers into the automatic transplanter.

Transferring the plugs to the pots or cell packs is becoming more and more automated. Nearly all the large greenhouse operations have automatic transplanters. The transplanters are programmed to lift the plugs from the trays and place the plugs in cell packs or pots filled with medium. Some of the transplanters cost more than $100,000. The speed at which they transplant makes them economical for large operations. Most models transplant between 10,000 and 43,000 plugs per hour, which is many more and much quicker than can be done by hand.

14–33. Technicians oversee the automated transplanter that removes the plugs from the trays and plants them in pots or cell packs.

14–34. This device sprays a bar code and an identifying label on each container as it exits the transplanter.

IRRIGATION

Irrigation or watering of bedding plants is the most important cultural practice in the greenhouse. Watering practices have a direct impact on the quality of the crop. Plants kept constantly moist tend to grow taller and have longer internodes than those allowed to dry out occasionally. Also, wet media reduces the exchange of gases around the roots, thus increasing the possibility of soil-borne diseases. On the other hand, plants allowed to become too dry may experience severe wilting that can stunt growth or cause death of tissues.

14–35. These workers set up the watering boom to fertigate a section of the greenhouse.

The frequency of watering is largely dependent on the weather and the size of the plants. Skilled greenhouse technicians closely monitor the crops in terms of watering needs. Sunny, hot days may require more than one watering. On cool, overcast days the plants may not need to be watered. When the crop is first transplanted, it should be watered immediately. However, it may not require more water for several days. Larger plants, nearing sale size, need more frequent waterings.

14–36. These geraniums in hanging baskets move from one side of the greenhouse to the other. As they do so, they are flooded with water.

When it is determined that water is needed, the crop is thoroughly watered. Thorough watering leaches soluble salts from the medium. Soluble salts are minerals and nutrients dissolved in the water. ***Leaching*** is a term used to describe the washing of salts from a medium as water passes through the medium. Without thorough watering, these salts build up in the medium and damage plant roots.

Automation has made watering easier in larger operations. The incorporation of automatic watering systems has permitted many more plants to be watered than could be by hand. One system involves overhead booms that sweep across the benches. Another has hanging baskets on a conveyor system. Each basket passes under a nozzle that floods the medium. Hanging baskets are also watered with spaghetti-tube irrigation systems. Some growers use the ebb and flood system of watering. Ebb and flood systems flood the benches with water. The water is absorbed into the medium from below. Even with the increased use of automation, greenhouse technicians must decide when to water. They must also hand-water plants missed by the automated system or plants that dry more quickly, such as the edges and corners of flats.

FERTILIZATION

Bedding plants respond best to a constant liquid feed beginning at the time of transplanting. This practice of fertilizing while irrigating is known as **fertigation**. Bedding plants require very light fertilizer applications relative to other greenhouse crops. The plants respond well to 200 ppm nitrogen, 100 ppm phosphorus, and 200 ppm potassium. Ageratum and marigolds, however, receive less fertilizer until they are showing color. Secondary macronutrients and micronutrients are added to the solution at much lower rates.

14–37. This person is mixing soluble fertilizer with water for fertigation.

Small growers often depend on packaged fertilizer mixes. These mixes are often formulated to provide all of the necessary macronutrients and micronutrients. Selection of fertilizers that contain the micronutrients makes it easier for the grower. The application of micronutrients is also especially important when the plants are grown in soilless mixes. Soilless mixes do not have the ability to hold nutrients the way clay and silt particles hold nutrients.

TEMPERATURE

Temperature is an important environmental factor in the production of bedding plants. Bedding plants like warm temperatures. Optimum growing temperatures for most

14–38. These bedding plants are exposed to cool day temperatures to slow plant elongation.

bedding plants range from 65 to 72°F. It is well known that bedding plants flower earlier and grow faster when given warm growing temperatures. However, warm temperatures can result in non-desirable features. The plants may stretch, develop weak stems, and have fewer bottom breaks.

Research has shown that growing bedding plants under cool day temperatures and warm night temperatures controls stretching of the plants. This technique is referred to as ***DIF*** or the difference between day and night temperatures. Keeping the night temperatures 6 to 8 degrees warmer than the day temperatures actually stops cell elongation. Keeping a cooler day temperature is sometimes difficult, especially in southern regions. This problem can be overcome by focusing on the first one or two hours of the day. If the temperature is kept cool the first hour or two of the day, the plant tends to accept that as the all-day temperature.

LIGHT

Most bedding plants grow best in full sunlight. One of the advantages of retractable roofs is the plants receive high light intensities. Even shade-tolerant plants, such as impatiens, wax begonia, and coleus, grow well under full sun in winter and spring. However, as production moves into late spring or early summer, some shading may be necessary.

GROWTH REGULATORS

Growers use growth regulators on bedding plants for several reasons. One reason is to improve the visual quality of the plants by keeping the plants short and stocky. Short and stocky plants are also easier to ship. In some cases, the application of growth regulators provides the side benefit of improved shelf life of plants. The two most widely used growth regulators for bedding plants are A-Rest and B-Nine. The use of growth regulators on vegetable plants is strictly forbidden.

Table 14–4. Recommended Use of Growth Retardants for Bedding Plants

Either A-Rest or B-Nine	A-Rest only	B-Nine only
Ageratum	Celosia	Petunia
Browallia	Cleome	
Dusty Miller	Coleus	
Dwarf Dahlia	Dianthus	
Geranium	Snapdragon	
Impatiens	Vinca	
Dwarf Marigold		
Tall Marigold		
Salvia		
Verbena		
Zinnia		

HANGING BASKETS

Hanging baskets are popular and as a result are being grown in greater numbers. Hanging basket production can be divided into two groups based on the plants used. Growers call them hard baskets and soft baskets. Bedding plants that are propagated by cuttings, such as fuchsia, geraniums, and New Guinea impatiens, make up the ***hard basket*** group. ***Soft baskets*** consist of bedding plants started from seed. Examples of soft basket plants are petunias, impatiens, browallia, gazania, portulaca, and thungbergia.

14–39. The pendulous habit of fuchsias make them ideal plants for hanging baskets. (Courtesy, *Greenhouse Product News*)

Production of hard baskets is a little more involved than with soft

14–40. Hand labor is required to hang and remove baskets.

baskets. Hard basket plants need to be started earlier, grown longer, and pinched during production to encourage branching. Hard baskets respond very well to DIF conditions. Soft baskets are relatively easy to grow. Soft baskets may only require 6 to 8 weeks to produce a final product versus 12 to 15 weeks for hard baskets. Large plugs are often used for soft baskets. Greenhouse growers maximize profits by growing large numbers of baskets over bedding plant flats.

14–41. Growing space is maximized in this house with thousands of fuchsia and New Guinea impatiens baskets being grown over bedding plant flats. Look closely to see the center basket receiving water as it rotates around the greenhouse on an automated rail system.

TONING BEDDING PLANTS

Up to the shipping or sale day, bedding plants have received warm temperatures, optimum fertility levels, and ample water to speed their growth and development. The place they are going, whether it is directly to the garden or to a sales display, is not so cozy. Therefore, the bedding plants must be prepared for a new environment. ***Toning*** describes the procedures used to prepare plants for a post-production environment.

Toning is accomplished by adjusting temperatures and fertility levels. Temperatures are lowered to tone bedding plants. Lower temperatures slow respiration, resulting in a buildup of carbohydrates in the plant tissues. The stored carbohydrates aid the plant in stressful situations associated with a change of environment. Dropping fertility levels also helps to prepare the plants. A good practice is to reduce fertilizer concentrations by 50 percent when flower buds have become visible.

Table 14–5. Cool Night Temperature Tolerances of Bedding Plants

Bedding plants toned at 50 to 55°F	Bedding plants toned at 58 to 62°F
Ageratum	Begonia
Alyssum	Celosia
Calendula	Coleus
Dianthus	Impatiens
Marigold	Pepper
Pansy	Tomato
Petunia	Vinca
Salvia	Zinnia
Snapdragon	

PERENNIALS

Many changes have swept the greenhouse industry. One of the more significant changes has been the surge in perennial production. ***Perennials*** are plants with life cycles of more than two years. Perennials offer consumers textures, colors, and growth habits of great interest. Plus, consumers see perennials as requiring less care than annuals. Perennials are planted once. They grow, go dormant, and re-emerge each spring.

Greenhouse growers have found perennials to be a good addition to their operations. Perennials have boosted sales. Perennials can be sold from early spring, before bedding plants, through the summer, and into autumn. In addition, hundreds of perennial species can be grown successfully alongside bedding plants.

Table 14–6. Value of Perennials Produced in the United States by Crop in 2001, USDA (in $1,000s)

Crop	Value in $1000s
Garden mums	103, 240
Hosta	39,021
Other perennials	345,483
Total	487,744

14–42. The trend has been to grow perennials in smaller pots such as this pulmonaria in a 4½ inch container. (Walters Gardens, Inc.)

14–43. Plant tags in these 4½ inch pots help to sell not-in-flower perennials. (Courtesy, Bell Nursery)

Propagation of perennials can be more difficult than most annuals. Requirements differ with each species. Many perennials have multiple seed dormancy mechanisms that must be satisfied for germination to occur. Perennials can also be propagated asexually. Tip cuttings, division, and tissue culture are common techniques used. Some perennials must be propagated asexually to remain true to form. Because of difficulty encountered with propagation, it is common for many growers to leave propagation to specialists and purchase rooted cuttings, liners, or plugs. The 128-size plug is popular.

Commercial growing mixes designed for greenhouse production work well for perennial production. A pH between 5.5 and 6.3 is good for most perennials. Rooted cuttings, liners, or plugs obtained from propagators are potted in pots from 3 to 5 inches in size. Some perennials, such as achillea, coreopsis, and veronica, can be finished in 6 weeks from plugs. Campanula, echinacea, and lavender are a few of many other species that can be finished in less than ten weeks from plugs. Astilbe, hosta, and daylilies can be finished in eight weeks or less from liners. Once big enough, the perennials can be transplanted to larger pots.

One of the differences between perennials and annuals involves flowering. Consumers are attracted to plants that are in flower. Given a choice, consumers select bedding plants that are in flower. It is easy to have annuals in flower. However, forcing perennials to flower can be challenging. Growers can do what is necessary to force some perennials to flower. An alternative is to have color pictures on plant tags to help sell not-in-flower species.

Various factors influence flowering of perennials. Requirements for forcing flowering are not fully understood for all perennial species. It is known that light duration, the age of the plant, and periods of cold are factors. Some species initiate flowers under short day conditions, while others initiate flowering under long day conditions. Day length can be manipulated with dark cloth or artificial lighting, much like it is controlled with potted flowering plant production. Some perennials must reach a certain level of maturity in order to flower. The number of leaves that have developed is a good indicator of the plant's developmental age. Some species require vernalization or a cold period to trigger flowering. Before shipping, propagators often provide materials with a cold treatment required for flowering.

Perennials are forced at night temperatures of 55 to 65°F. Growers often maintain temperatures closer to 65°F for the first few weeks to get the plants well established. Then, they gradually drop the temperatures to 55°F to slow growth. Negative DIF (cool days, warm nights) has proven an effective way to keep plants compact. Growth regulators can also be used in some cases for this purpose.

Nutritional requirements for most perennials are light. A constant liquid feeding of 50–100 ppm of nitrogen with a balanced fertilizer is generally sufficient. Micronutrients should be added if lacking from the water-soluble fertilizer.

Greenhouse grown perennials are subject to the same pests as other greenhouse crops. A good integrated pest management program can eliminate many potential problems.

REVIEWING

MAIN IDEAS

Bedding plants consist of annual plants valued for their flowers, or attractive foliage, and vegetables and herbs. Sales of bedding plants have grown significantly. The increased use of automated systems and larger greenhouse operations has significantly increased bedding plant production.

Production of bedding plants begins, in most cases, with sexual propagation or the use of seed. Growers schedule the sowing of seed to meet sales dates. Seed is carefully stored to maintain viable seed and to ensure high germination rates. Small growers rely on hand seeding, while larger operations have automatic seeders. The grower must be aware of the requirements of water, light, and temperature for successful germination of seed.

Plugs have revolutionized the bedding plant industry. Roughly two-thirds of all bedding plants are grown as plugs. Seeds are planted in individual cells in a plastic tray that might hold between 70 to 800 plants. Commonly used growing media have mostly peat moss, with some vermiculite, perlite, or soil added. The goal of the grower is to produce plugs that are uniform, have short internodes, and have a high number of dark green leaves. The entire process in producing quality plugs takes 6 to 10 weeks. The plugs are sold and shipped to operations around the country or transplanted to larger containers for finishing.

The industry trend has been to finish bedding plants in 4-inch pots or cell packs with no more than 48 plants per flat. The larger containers extend the shelf life of bedding plants. Plugs in large operations are automatically transplanted from the plug trays to the pots or cell packs. Irrigation or watering has the greatest impact on the quality of the crop. Most growers employ the practice of fertigation. Bedding plants like warm temperatures. Optimum growing temperatures for most bedding plants range from 65 to 72°F. Stretching of bedding plants can be reduced by growing bedding plants under cool day temperatures and warm night temperatures. This technique is referred to as DIF or the difference between day and night temperatures. Most bedding plants grow best under full light conditions, with shading only necessary in late spring and early summer. Hanging baskets are divided into two groups, hard baskets and soft baskets. Before plants leave the greenhouse for sale, they are toned to prepare them for a change of environment and to extend their shelf life.

Perennials can be sold from early spring, before bedding plants, through the summer, and into autumn. Because of difficulties encountered with propagation, many growers purchase rooted cuttings, liners, or plugs. Commercial growing mixes with a pH between 5.5 and 6.3 work well for perennial production. Forcing perennials to flower can be challenging, and growers must understand the various factors that influence flowering of perennials. Perennials are forced at night temperatures of 55 to 65°F. A constant liquid feeding of 50 to 100 ppm of nitrogen with a balanced fertilizer is generally sufficient.

QUESTIONS

Answer the following questions using complete sentences and correct spelling.

1. What have been the recent trends in the bedding plant industry?
2. How are bedding plants scheduled?
3. What characteristics do quality seed possess?
4. What factors influence seed germination?
5. How have plugs revolutionized the bedding plant industry?
6. Why is irrigation considered to be so important?
7. How does the technique DIF influence bedding plant growth?
8. Why are growth regulators used with bedding plants?
9. How are bedding plants toned?
10. What are some of the cultural requirements for producing perennials?

EVALUATING

Match the term with the correct definition. Write the letter of the term on the line provided.

a. bedding plants
b. DIF
c. cell pack
d. fertigation
e. shelf life
f. break
g. hard baskets
h. finished
i. toning
j. plugs

______1. Refers to the development of lateral branches that contribute to a bushy appearance.

______2. Bedding plants, such as fuchsia, geraniums, and New Guinea impatiens, that are propagated by cuttings and used in hanging baskets.

______3. A term used to describe procedures that prepare plants for a post-production environment.

______4. Molded plastic containers divided into two, three, four, or six separate growing compartments.

______5. The practice of fertilizing while irrigating.

______6. The final pots or packs in which bedding plants will be grown to saleable size.

______7. Technique used that involves the difference between day and night temperatures.

______8. Include herbaceous annual plants used for ornamental display or vegetable production.

______9. Refers to the time the plant maintains its health while on display for sale.

_____10. Small plants grown in a small amount of medium in divided trays.

EXPLORING

1. Use the knowledge gained from this chapter to schedule, start, and grow bedding plants at home. Transplant the plants to the home landscape when weather permits.

2. Visit a local greenhouse in the spring. Observe their growing techniques, the plants they produce, and the supplies they use. Record your observations.

3. Ask your instructor to allow you to conduct an independent study project in the school greenhouse or to be a greenhouse assistant. Keep records on your experiences and, if applicable, maintain a supervised agricultural experience program.

CHAPTER 15

Pest and Disease Management

OBJECTIVES

This chapter is about plant pest and disease management. It has the following objectives:

1. Describe plant health and causes of plant disorder.
2. Define integrated pest management.
3. Explain how crops can be monitored for pest and disease problems.
4. Identify major pests in the greenhouse.
5. Identify major plant diseases in the greenhouse.
6. Describe methods of control for the major greenhouse pests.
7. Demonstrate methods of safe chemical pesticide application.

TERMS

aerosol
aphids
bactericide
biological control
botanical insecticides
Botrytis blight
chemical control
cultural/physical control
damping-off
dip
disease
drench
Erwinia
fumigant
fungicide
fungus gnats
granular
greenhouse sanitation
herbicide
horticultural oils
infectious disease
insect growth regulators
insecticidal soaps
insecticide
integrated pest management (IPM)
leaf miners
mealybugs
mites
miticide
noninfectious disease
pathogens
pest
pesticide
powdery mildew
Pythium
Phytophthora
plant health
Rhizoctonia
scale
spray
Thielaviopsis
virus
western flower thrips
whitefly
worms
yellow sticky traps

15–1. Growers need to be on the lookout for pests and diseases.

NO PROBLEMS reduce the value of floriculture products more than pests and diseases. Who wants to buy a plant covered with aphids, a plant with spider mite webs, or a plant with stem rot? People will buy lopsided plants. They will buy plants with broken stems. But they will not touch plants with visible pest problems.

Growers dislike pests for different reasons. They know what a quality plant should look like and strive to produce the highest quality possible. They also know the damage pests and diseases can do to growth and development. Unless they control pests and diseases, their crops lose value and the business loses potential revenue.

MANAGING PLANT HEALTH

In the process of producing greenhouse crops, growers must address plant health. ***Plant health*** refers to plants that are free of pests and disease. Plant health is important during the production of the crop. It is also important that plants continue to be healthy after they are sold. Healthy plants have clean foliage and flowers, and they have a good rate of growth.

15–2. Healthy plants are free of pests and diseases.

Pests and diseases can jeopardize plant health. A ***pest*** is a living organism that can cause injury or loss to a plant. Common greenhouse pests include insects, mites, bacteria, fungi, viruses, and weeds. Some pests can be the cause of disease. ***Disease*** is defined as a disturbance to the normal, healthy growth and development of a plant. Diseases are generally classified as being infectious or noninfectious. Bacteria, fungi, or virus can cause ***infectious disease***. These organisms are often referred to as disease ***pathogens***. An infectious disease can be spread to other plants. ***Noninfectious diseases*** are caused by environmental imbalances and cannot be spread to other plants. Examples of noninfectious disease include over watering and air pollution damage.

Plants are most susceptible to disease when they are under some type of stress. The stress is usually associated with environmental factors. While the environmental condition itself may or may not cause a noninfectious disease, the stress it creates can reduce a plant's ability to fight off infectious disease. For instance, over watering alone can cause the death of root tissues, but it also weakens the plant's defenses against soil-borne disease organisms.

Growers have control over many environmental factors that can keep a plant healthy. Care can be taken to provide growing medium with desired drainage, aeration, and pH.

15–3. Proper watering practices reduce disease problems.

Plants can be planted at the proper planting depth. Optimum nutrient levels can be maintained with fertilizers. One of the most important factors to control is recommended watering practices. Also, temperature, light intensity, and air quality can be adjusted to meet the needs of the specific crop.

No matter how well crops are grown, pests and diseases will become problems from time to time. The very nature of greenhouse crop production leads to some disease problems. In most cases, crops are of the same species, variety, and cultivar. Being of identical genetic makeup, they are vulnerable to infectious disease. An infectious disease can easily spread from one plant to another. Greenhouses tend to be humid. Humid conditions are ideal for many fungal diseases. Successful growers produce crops with a plan for dealing with pests and diseases.

15–4. Computer programs have been designed to assist with integrated pest management. (Courtesy, Agricultural Research Service, USDA, Jack Dykinga)

INTEGRATED PEST MANAGEMENT

Control of pests and diseases in the modern greenhouse is accomplished with well thought-out strategies. Use of a wide range of strategies to control pests and diseases is known as ***integrated pest management (IPM)***.

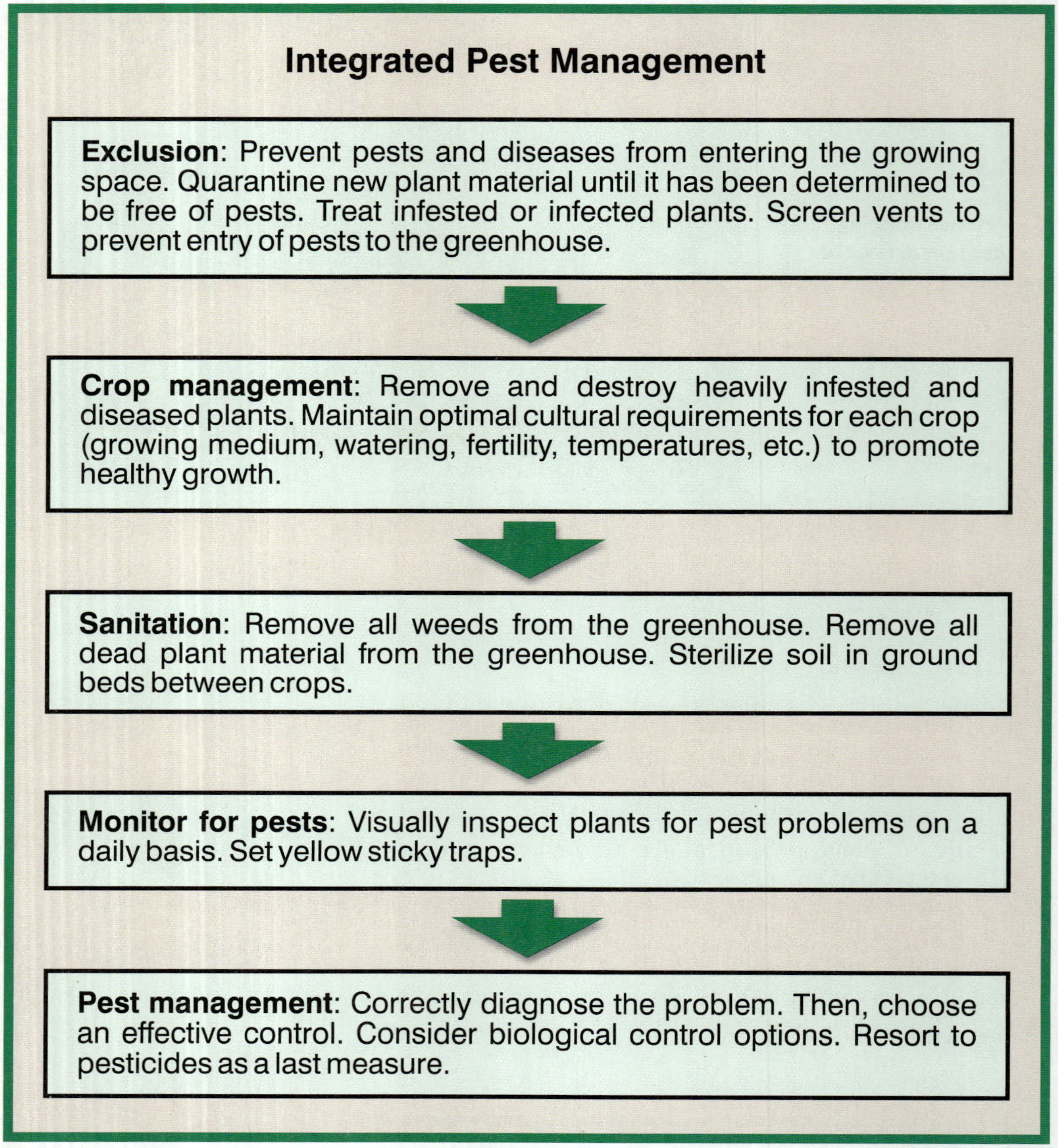

15–5. An IPM flow chart.

MONITORING FOR PESTS

The key to successful IPM programs is the careful monitoring for pests. This involves casual observation while performing other tasks as well as scouting for pest problems on a weekly schedule. Weekly inspections should be made by randomly selecting plants at the middle and sides of benches. The number of plants checked should total a minimum of 1 percent of the crop during each study. Thoroughly check the plants by looking at the top, middle, and bottom leaves of plants. Often pests are located on the undersides of leaves, so be sure to look there.

Yellow sticky traps are effective in monitoring pest populations. Brightly colored, they attract whitefly, western flower thrips, aphids, leaf minors, and fungus gnats. The insects land on the traps, get stuck, and die. The sticky traps are useful as general indicators of a pest population. They also help provide early detection of insect problems. Early detection gives the grower time to prepare an IPM program. Sticky cards should be put in horizontal positions just below the tops of the crops. All the cards in a house should face the same direction, too.

15–6. Yellow sticky traps are an effective means to monitor insect pest populations.

MAJOR GREENHOUSE PESTS AND DISEASES

Correct identification of the pest or disease cannot be understated. How can a pest be controlled if it is unknown? Unless the pest is known, proper treatment cannot be administered. For instance, a doctor would not prescribe cold medicine to treat athlete's feet. Following are a number of the more common pests and diseases found in the greenhouse:

Pests

Aphids—Aphids are pear-shaped, soft-bodied, usually wingless insects. They are often green or yellowish in color. Aphids have the ability to reproduce very rapidly. Astonishingly, they give birth to live young that are pregnant! Aphids use their mouthparts to pierce the plant and suck out juices. Aphids attack a wide variety of greenhouse plants.

15–7. Ladybugs feed on smaller yellow-green aphids. (Courtesy, Agricultural Research Service, USDA)

Fungus gnats—Fungus gnats are long-legged, winged, gray-black insects less than an eighth of an inch long. The larvae of fungus gnats feed on root hairs and tunnel into plant stems. They prefer growing media that is constantly moist.

Leaf miners—Leaf miners are small stocky flies. The adult deposits eggs inside a leaf. The eggs hatch and the larva feeds on the interior of the leaf, making tunnels as it moves along. Chrysanthemums are subject to leaf miner damage.

15–8. Shown are a fungus gnat larva and adult. (Courtesy, Fredric Miller and Phil Nixon, University of Illinois Extension)

Mealybugs—Mealybugs are slow-moving, oval-shaped, whitish insects. They have a waxy finish and produce small cottony masses. Mealybugs pierce plant leaves and suck the plant juices. As with aphids, they give birth to living nymph.

Mites—Mites are not insects. They have eight legs and are related to spiders. Mites pierce plant leaf tissues and suck juice. Symptoms include a yellow speckled appearance to the leaf, and in severe cases, yellowing of the leaves and defoliation.

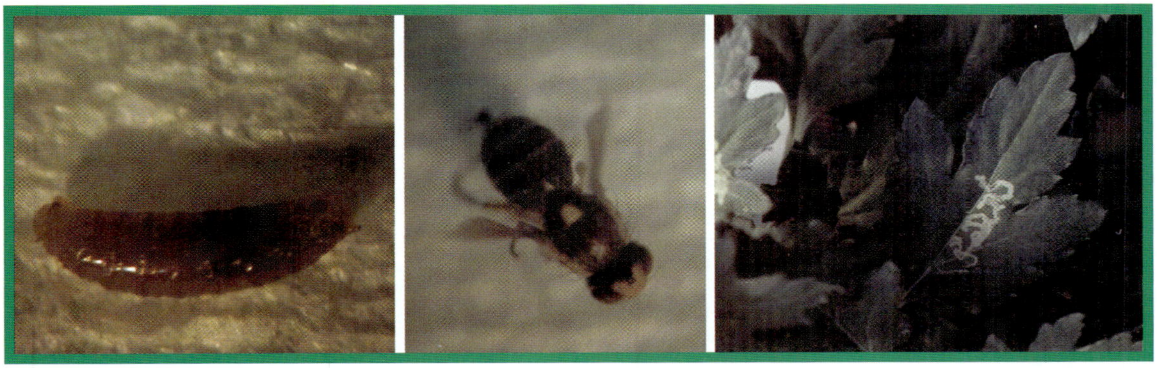

15–9. Leaf miner larva, adult, and damage caused to a chrysanthemum. (Courtesy, Fredric Miller and Phil Nixon, University of Illinois Extension)

Two-spotted or red spider mites are among the most serious greenhouse pests. Two-spotted mites have two dark spots on their back and appear as tiny specks. Cyclamen mites infest a broad range of plants and are not visible to the naked eye. Bulb mites damage lily bulbs and the developing shoots. Spider mites create a complex network of webbing as they move about a plant.

15–10. Mealybugs are often found in the axils of leaves. (Courtesy, Fredric Miller and Phil Nixon, University of Illinois Extension)

Scale—Many types of scale insects infest greenhouse plants. Typically, they have flat, oval, often

15–11. Spider mites on an acacia and a Ficus. Notice the speckled appearance of the leaves caused by feeding. (Courtesy, Fredric Miller and Phil Nixon, University of Illinois Extension)

15–12. Photographs of scale on a Ficus. (Courtesy, Fredric Miller and Phil Nixon, University of Illinois Extension)

brown bodies. They may or may not be covered with an armored shell. Scale insects pierce plant leaves and stems and suck juices.

Western flower thrips—Western flower thrips are small insects with two pairs of fringed wings. They are dark brown in color. Western flower thrips have rasping mouthparts that scrape plant tissue. The damage they cause to many kinds of plants often appears as whitish discoloration.

Whiteflies—Whiteflies are small insects that are not, surprisingly, white. They generally camp out on the undersides of leaves where they pierce the tissues and suck juices. Their flat, scale-like larvae feed on the undersides of the leaves. The whitefly is a major pest of poinsettias and fuchsias.

Worms or caterpillars—Caterpillars are the larva of various moth species. They damage greenhouse crops by eating the plants.

15–13. A thrip. (Photo by M. Herbut, Courtesy of Applied Bio-monics, Ltd., contributed by Mike Cherim, The Green Spot, Ltd.)

15–14. Shown is a whitefly greatly magnified. (Courtesy, Agricultural Research Service, USDA)

Diseases

Botrytis blight—*Botrytis* blight is a fungal disease that can attack nearly all greenhouse crops, and is a common problem with cut flower storage. It causes a brown rotting and develops fuzzy, gray mold as it produces spores. *Botrytis* is most common when temperatures are between 60 and 70°F, air circulation is poor, and humidity is high.

Erwinia—*Erwinia* is a bacterial disease that causes rotting of plant tissues. The bacteria enter the plant through wounds. It is a common problem on *Dieffenbachia, Philodendron*, chrysanthemum, and cyclamen.

15–15. *Botrytis* on an impatiens. (Courtesy, James E. Schuster)

15–16. *Botrytis* is a serious fungal disease, shown here on a dahlia. (Courtesy, James E. Schuster)

Powdery mildew—Powdery mildew is a fungal disease. Characteristic symptoms include a white dusty coating to leaves and flowers. Infected plants can become stunted. High humidity contributes to powdery mildew problems.

Pythium—*Pythium* is a fungal disease that attacks greenhouse plants under cool, wet conditions. It is one fungus that can cause damping-off. **Damping-off** is a term used to describe the early death of seedlings that have germinated. More mature plants can develop root and stem rots as a result of *Pythium*, particularly when growing medium has poor aeration.

Phytophthora—*Phytophthora* is a fungal disease similar to Pythium. It causes crown and stem rots in cool, wet conditions.

15–17. Powdery mildew can be a problem with roses. (Courtesy, James E. Schuster)

15–18. Damping-off is often caused by *Pythium*. (Courtesy, James E. Schuster)

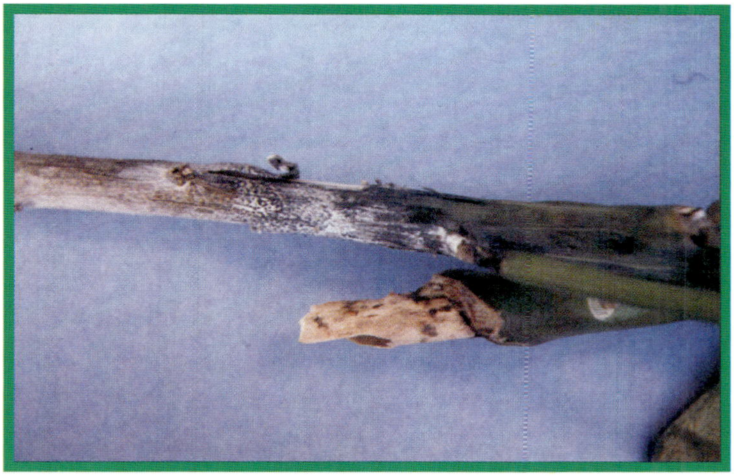

15–19. Stem rot is shown in this photograph. (Courtesy, James E. Schuster)

Rhizoctonia—*Rhizoctonia* is a fungal disease prevalent under wet and warm conditions. It is a common disease problem in the southeastern region of the country. *Rhizoctonia* causes damping-off, as well as root and stem rots.

Thielaviopsis—*Thielaviopsis* is a fungus that causes root and stem rots. Problems caused by *Thielaviopsis* are favored by cool, moist conditions.

Virus—Tobacco mosaic and aster yellows are two common viral diseases associated with greenhouse crops. Symptoms of infection include discoloration of plant tissues, stunting of growth, and deformed growth. The spread from one plant to another is primarily by feeding greenhouse insects. Tobacco mosaic virus can be spread to plants from the hands of workers who smoke.

15–20. The streaking of this tulip is caused by a virus. (Courtesy, James E. Schuster)

15–21. Aster yellows, a virus disease, has caused discoloration of the foliage of this marigold. (Courtesy, James E. Schuster)

PEST CONTROL STRATEGIES

For successful management of pests, the IPM program must be a year-round program. Also, IPM control measures for a specific crop, poinsettias for example, should begin before the plants enter the greenhouse. The strength of IPM is the combination of control measures used. Four broad areas of control include sanitation, cultural/physical control, biological control, and chemical control.

Greenhouse Sanitation

Many pest problems can be greatly reduced, if not eliminated, with greenhouse sanitation. ***Greenhouse sanitation*** is simply the efforts made to keep a greenhouse clean.

5–22. Concrete floors, as in this greenhouse, are easy to keep clean.

15–23. Weeds are controlled in this greenhouse with a fabric weed barrier.

15–24. Diseased or infested plants should be removed from the greenhouse and disposed.

Many modern greenhouse ranges have been built with concrete floors partly because they are easier to keep clean than gravel floors.

One aspect of greenhouse sanitation involves the removal of weeds from the interior of the greenhouse and the immediate area outside the greenhouse. The importance of weed control cannot be understated. Weeds harbor pests. Control measures applied to a crop fail to control pests that find safety on the weeds. After awhile, the pests migrate to the crop to cause damage. Only a few herbicides are labeled for use in the greenhouse.

Another important sanitation practice is the removal of plant debris and other debris from floors and benches. Debris often houses disease organisms and pests. Severely infested or infected plants should also be removed and disposed of properly.

Cultural/Physical Control

Cultural/physical control methods are those methods that physically prevent activities of pests. Used alone they probably will not provide complete control of pests. However, they can significantly reduce certain problems. Cultural/physical controls are also safe to humans and relatively easy to implement.

Stop the introduction of pests to the greenhouse when possible. Inspect all plants thoroughly before moving them into the greenhouse. Accept only clean plants purchased from reputable suppliers. Admitting infested plants to the greenhouse usually leads to serious control problems.

15–25. Good air circulation reduces fungal disease problems.

Maintain the ideal growing environment for the crops. Plants that are healthy are better able to fight disease organisms and pests. A good analogy is people and colds. People who eat right, exercise, get plenty of rest, and are under little stress are more resistant to colds. People who get run down, are out of shape, and eat a poor diet are more susceptible to colds.

Fungal diseases can be reduced by providing good air circulation around the plants. Fungi rely on moisture to grow and spread. Air circulation helps to keep plant leaves dry.

One physical control method that has gained in popularity is the use of screens over greenhouse openings. The fine mesh of the screens prevents pests from flying into the greenhouse environment. The one drawback with insect screening is it can restrict the airflow somewhat.

The yellow sticky traps used as monitoring tools also serve as a means of physical control. Flying insect pests attracted to sticky traps fly onto the trap and get stuck. In a short while, they die.

15–26. Insect screens over greenhouse openings can help keep harmful insects out. (Courtesy, Ludvig Svensson, Inc.)

Biological Controls

Biological controls involve the use of living organisms to control pests. They may be microbial organisms, parasitic organisms, or predators. Biological control organisms for greenhouse use are found in nature and are considered environmentally safe.

15–27. A parasitoid wasp, *Aphidius* species, laying eggs in the abdomen of aphids. (Photo by M. Badgley, courtesy of Buena Biosystems, Inc., contributed by M. Cherim, The Green Spot, Ltd)

Microbial organisms include bacteria and fungi. A bacteria, *Bacillus thurengiensis,* effectively controls caterpillars. Aphids and whitefly can be controlled to an extent by species of bacteria and fungi. The bacteria and fungi are natural diseases of those insects. However, high humidity in the greenhouse required for the beneficial fungi also promotes the growth of disease-causing fungi.

Parasitic organisms help to control some pests. The parasites are natural enemies of the pest and live off the pest organism. An example is a tiny parasitic wasp that lays its eggs on

15–28. A parasitoid wasp, *Encarsia formosa.* It lays eggs in the larva of whitefly. (Photo by B. Costello, courtesy of Applied Bio-nomics, Ltd., contributed by M. Cherim, The Green Spot, Ltd)

15–29. The big-eyed bug is a predator with mouthparts that allow it to inject white fly larvae with a digestive enzyme. It then sucks out their insides. (Courtesy, Agricultural Research Service, USDA)

the whitefly larva that feeds on plant leaves. The eggs hatch with the wasp larva inside the whitefly larva. The wasp larva proceeds to eat the whitefly larva. The wasp matures, emerges from what is left of the whitefly, mates, and looks for whitefly larva on which to lay the next generation of eggs. There are some restrictions with the use of parasitic insects. Chemical pesticides cannot be used in the greenhouse without killing the beneficial parasites. The parasites are also not effective with heavy infestations of plant pests.

Predatory organisms can be released in the greenhouse to devour certain plant pests. A beetle attacks whitefly larva and adults. A mite is used to control thrips. Ladybugs eat aphids. As with parasitic organisms, chemical pesticides should not be used with predatory organisms. Also, predatory and parasitic organisms should be released when pest populations are small.

Chemical Control

The use of chemicals to control pests and diseases is ***chemical control***. The chemicals used are called ***pesticides***. Although once used almost exclusively, control of pests with the use of pesticides is now viewed as only one component of an IPM program. In fact, use of chemical pesticides is now often done only when it is absolutely necessary. Application of pesticides must be done safely to reduce potential injury to people and the environment.

Different types of pesticides are used to control different pests. ***Insecticides*** are used to control insects. Mites are controlled by ***miticides***. A grower would use a ***fungicide*** to control fungi, ***bactericide*** for bacteria, and ***herbicide*** for weeds.

Pest control has been made safer with the introduction of less toxic pesticides. This group of "soft" pesticides includes botanical insecticides, horticultural oils, insect growth regulators, and insecticidal soaps. ***Botanical insecticides*** are natural materials obtained from plants that are toxic to the insect pest. ***Horticultural oils*** are sprayed on the insects and clog the breathing pores of the insect, causing suffocation. ***Insect growth regulators*** disrupt the growth and development of the insects. ***Insecticidal soaps*** dissolve the protective membranes of insects, bringing on death.

There are several methods used to apply pesticides in the greenhouse. The key is to get the pesticide to the pest that is to be controlled. The methods vary with the formulation of the chemical.

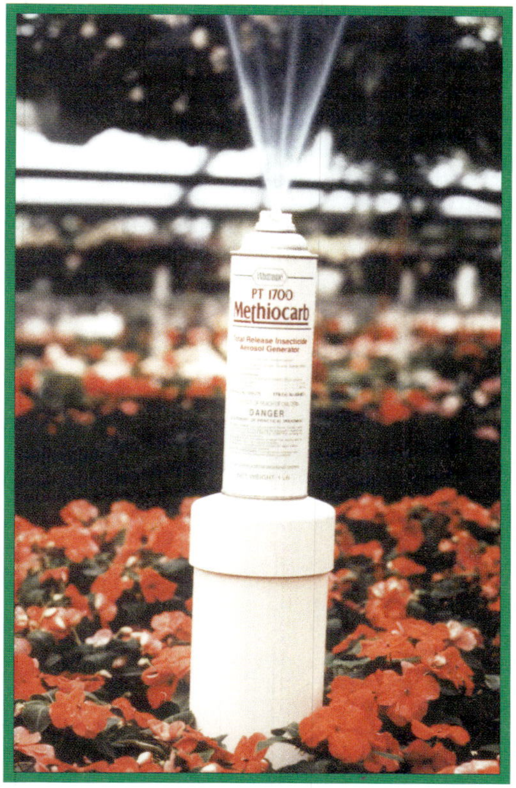

15–30. Aerosols release pesticides into the air. (Courtesy, *Greenhouse Product News*)

Aerosols—Aerosols consist of atomized spray or smoke particles distributed through the air. Cans containing the pesticide can be set to release the aerosol that creates a fog in the greenhouse. Once started, the applicator can safely leave.

Dip—Plants can be dipped or submerged in a pesticide that has been mixed with water. Easter lily bulbs are dipped in a miticide prior to planting to control bulb mites.

Drench—Some materials are used as a drench. A common practice is to drench poinsettias with a fungicide to control soil-borne disease. The material, mixed with water, is simply poured into the pots.

Fumigants—Fumigants are poisonous gases that are distributed through the air to all parts of the greenhouse. They are particularly effective with flying insects and pests exposed to the gas.

Granular—Some pesticides are in a granular or pellet form that is carefully measured and applied on the surface of the growing medium. These pesticides are referred to as systemic because the plant's roots absorb them into their systems.

Spray—Pesticides may be sprayed on the plant. The small droplets of the spray cover the surface of the plant. Care must be taken with sprays to cover parts of the plant where the pest is found, such as the underside of the leaves.

15–31. A measured amount of granular pesticide is placed in poinsettia pots.

SAFE USE OF PESTICIDES

There is potential danger when pesticides are used. The applicator can be poisoned. Plants can be damaged. The environment can be polluted and ground water contaminated. Therefore, caution must be practiced whenever pesticides are used.

Some rules to observe when applying pesticides follow:

- Use pesticides only when needed.
- Select the least toxic pesticide that is labeled to control the pest.
- **Read the label of the pesticide carefully.**
- Wear protective clothing when mixing and applying pesticides.
- Apply pesticides when the weather is good (not too hot or windy).
- Apply the amount that is called for on the pesticide label.
- Apply pesticides so complete coverage of where the pests are located is provided.
- Use equipment that is properly calibrated.
- Label holding tanks indicating what type of pesticide has been used.
- Dispose of old pesticides and empty pesticide containers properly.
- Label crops that have been treated and note the safe time for workers to re-enter the area.

15–32. This person is using an electrostatic sprayer to apply fungicide to a poinsettia crop. (Courtesy, Agricultural Research Service, USDA)

15–33. The most important time spent in pesticide application is the time reading the label.

15–34. This person is shown preparing a fungicide for application.

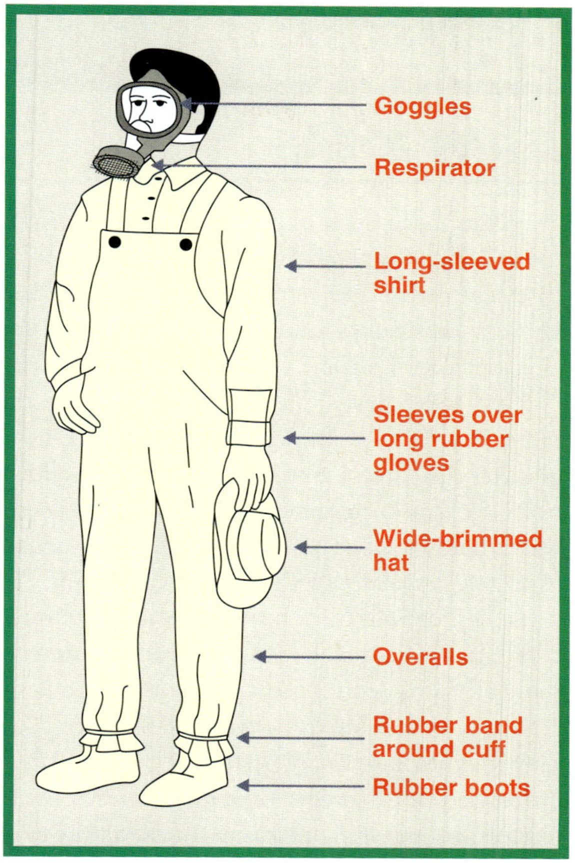

15–35. Protective clothing must be worn when working with pesticides.

15–36. Application equipment must be clearly marked as to its use and treated crop areas properly labeled to remind workers of pesticide application.

REVIEWING

MAIN IDEAS

Plants that are free of pests and disease are important during the production of the crop and after they are sold. Healthy plants have clean foliage and flowers, and they have a good rate of growth. Pests and diseases can jeopardize plant health. Common greenhouse pests include insects, mites, bacteria, fungi, viruses, and weeds. Pests can be the cause of disease or a disturbance to the normal, healthy growth and development of a plant. Diseases are generally classified as being infectious or noninfectious.

Control of pests and disease problems in the modern greenhouse is accomplished with a wide range of strategies known as integrated pest management (IPM). The key to successful IPM programs is the careful monitoring for pests. Yellow sticky traps are effective in monitoring pest populations. Common greenhouse pests include aphids, fungus gnats, leaf miners, mealybugs, mites, scale, western flower thrips, whitefly, and worms or caterpillars. Common disease pathogens include *Botrytis* blight, *Erwinia*, powdery mildew, *Pythium*, *Phytophthora*, *Rhizoctonia*, *Thielaviopsis*, and virus.

IPM programs must be year-round programs. Four broad areas of control include sanitation, cultural/physical control, biological control, and chemical control. Many pest problems can be greatly reduced with greenhouse sanitation, which is simply keeping a greenhouse clean. Cultural/physical control methods are those methods that physically prevent activities of pests. Examples include stopping the introduction of pests to the greenhouse, maintaining the ideal growing environment for the crops, and using screens over greenhouse openings. Biological controls involve the use of microbial organisms, parasitic organisms, or predators to control pests. Chemical control of pests and diseases involves chemicals called pesticides.

Different types of pesticides are used to control different pests. Insecticides control insects. Mites are controlled by miticides. A grower would use a fungicide to control fungi, bactericide for bacteria, and herbicide for weeds. A group of safer materials known as "soft" pesticides include botanical insecticides, horticultural oils, insect growth regulators, and insecticidal soaps. Methods used to apply pesticides in the greenhouse include aerosols, dips, drenches, fumigants, granules, and sprays. There is potential danger when pesticides are used. Therefore, caution must be practiced whenever pesticides are used.

QUESTIONS

Answer the following questions using complete sentences and correct spelling.

1. What is meant by plant health?
2. How do infectious and noninfectious diseases differ?
3. What can be done to help a plant fight off pests and diseases?
4. Why are crops monitored for pests and diseases?
5. What are some common greenhouse pests?
6. What are some common greenhouse pathogens?
7. Why is integrated pest management effective?

8. What are four control methods applied in an IPM program?
9. Why are pesticides often considered to be a last resort?
10. Why is safety so important when using pesticides?

EVALUATING

Match the term with the correct definition. Write the letter of the term on the line provided.

a. aphids
b. *Botrytis*
c. drench
d. fungicide
e. granular
f. greenhouse sanitation
g. horticultural oils
h. integrated pest management (IPM)
i. mites
j. pest
k. *Pythium*
l. yellow sticky traps

_____1. A pesticide, mixed with water, is simply poured into the pots.
_____2. A living organism that can cause injury or loss to a plant.
_____3. The efforts made to keep a greenhouse clean.
_____4. Use of a wide range of strategies to control pests and diseases.
_____5. Clogs the breathing pores of insects on which it is sprayed, causing suffocation.
_____6. A pellet-form pesticide that is carefully measured and applied on the surface of the growing medium.
_____7. A pesticide that controls fungi.
_____8. A fungal disease that can attack nearly all greenhouse crops; it causes a brown rotting and develops fuzzy, gray mold as it produces spores.
_____9. Pear-shaped, soft-bodied, usually wingless insects that are often green or yellowish in color.
_____10. A fungal disease that attacks greenhouse plants under cool, wet conditions and may cause damping-off.
_____11. Effective in monitoring pest populations.
_____12. Pests with eight legs and related to spiders.

EXPLORING

1. Set up your own scouting activity at home or in the school greenhouse. Monitor the plants for pest and disease problems. Record your observations. Suggest ways in which the problems can be managed.

2. Study the safe use of pesticides in greater depth than is presented in this chapter. Obtain study manuals from your state extension office. Attend training sessions if they are offered. Once you feel competent, test your knowledge by taking the state-licensing exam on pesticide use.

Glossary

Abscisic acid—A growth-inhibiting hormone largely responsible for seed dormancy and contributes to senescence along with ethylene.

Acclimatization—The process in which a plant becomes accustomed to a new environment.

Acidic—Refers to a pH or relative concentration of hydrogen ions in a solution of 1 through 6.

Acrylic structured sheets—Glazing material manufactured with twin walls in the same fashion as polycarbonate sheets. They have a high light transmission of 86%, second only to glass. They are less flexible than polycarbonates and they are more prone to hail damage.

Adventitious roots—Roots that begin growth from the stem or a leaf.

Advertising—Efforts to encourage sales and that create an image.

Aeration—The exchange of gases in the media.

Aerosol—Atomized spray or smoke particles distributed through the air.

African violet—Potted flowering plant native to areas south of the equator on the African continent. Flower colors fall in the white-pink-purple range. They are produced for Valentine's Day, Easter, and Mother's Day.

Alkaline—A pH of 8 through 14.

Alstroemeria—Known as the Peruvian lily or Inca lily, it is everblooming, produces high yields, and it is easy to grow under cool temperatures. Flowers of the *Alstroemeria* have a long vase life, two weeks or more, and come in many colors including yellow, orange, red, pink, purple, lavender, white, and bicolor.

Amaryllis—Tropical or sub-tropical bulb found growing naturally in South America. Varieties of amaryllis have been improved through hybridization, providing a wide range of large, colorful flowers.

Analog controls—Environmental controls that use proportioning thermostats or electronic sensors to run amplifiers and electronic circuitry. They integrate the operations of heating and cooling equipment.

Annuals—Herbaceous plants that grow from seed to flower in one season before dying.

Anther—The part of the stamen that produces and holds the pollen.

Anthocynin—A plant pigment responsible for many of the blue, purple, and red colors.

Aphids—Pear-shaped, soft-bodied, usually wingless insects, green or yellowish in color, and have the ability to reproduce very rapidly.

Apical dominance—The term to describe the apical bud's dominance over the lateral buds.

Apical meristem—The primary growing point of the stem.

Arm bouquet—A grouping of flowers and foliages that is tied together and cradled in the bride's left arm.

Asexual propagation—Plant reproduction using vegetative parts of a plant.

Asiatic lilies—Bulb crop native to the Asian Continent. They come in a wide range of colors, including yellow, white, red, and orange. They are forced throughout the year.

Attitude—One's outlook on life.

Automated seeders—Automated system that permits the sowing of hundreds of flats of bedding plant seeds.

Automated transplanters—Automated system in which small plants are removed from seed flats and planted into pots for finishing.

Auxins—A group of compounds that stimulate certain plant responses. The major plant response influenced by auxins is cell elongation. Auxins are produced in the apical meristem of a plant's stem and migrate down the stem.

Available water—The water that can be absorbed by the plant's roots.

Bark—Organic growing media component that improves the moisture holding ability and aeration of a mix. Bark ground into fine pieces is the most useful because of its uniform qualities.

Basal plate—A hardened portion of stem tissue to which bulb scales are attached. Roots develop along the outside edge of the basal plate.

Basal rooters—Plants whose roots form at the base of the stem cutting.

Bedding and garden plants—Herbaceous, annual flowers and vegetables that lack the ability to survive freezing temperatures. Bedding plants include impatiens, petunias, marigolds, tomatoes, and many other plants.

Bench—A structure that holds crops off the ground.

Biological control—The use of living organisms to control pests.

Botanical insecticides—Natural materials obtained from plants that are toxic to the insect pest.

Botrytis blight—A fungus disease that can attack nearly all greenhouse crops and is a common problem with cut flower storage. It causes a brown rotting and develops fuzzy, gray mold as it produces spores.

Bract—Modified leaves that take on a petal-like appearance.

Breaks—Refers to the development of lateral branches that contribute to a bushy appearance.

Broker—An agent who purchases flowers from growers and sells the materials to both wholesale and retail florists.

Bud scale—Cover and protect undeveloped leaves, stem, or flowers during dormancy.

Budding—A form of grafting in which the scion consists of a bud.

Buds—Structures that contain undeveloped leaves, stem, or flowers.

Bulb—Short, flattened stems that bear fleshy food storage leaves. Examples of bulbs include tulip and Narcissus.

Bulbets—Tiny bulbs produced below the ground.

Bulbils—Tiny above ground bulbs produced in the axils of leaves.

Bulk density—The ratio of the mass of dry solids in a media to the volume of the media.

Business plan—An organizational business tool to state the business' purpose and goals and estimate the finances needed to fund a new flower shop.

Buyer—Person who locates sources of products and places bids or offers for purchase.

Calcined clay—Growing media component in which clay aggregates are heated to form hard, stable particles. Calcined clay improves water drainage and aeration in a media, has a high bulk density, and also has some cation exchange capacity.

Callous—A group of cells with no particular function.

Calyx—The collection of the sepals of a flower.

Cambium—A layer of cells where cell division takes place. These dividing cells become either xylem or phloem cells depending on which side of the cambium they are located.

Capillary mats—Porous mats laid on a bench and wetted. The media in the pots absorb the water by capillary action.

Career success—Involves those things that will help a person succeed in the world of work.

Carotene—A pigment that produces orange-yellow colors. It plays a much smaller role in photosynthesis than chlorophyll. Carrots have a high carotene content.

Case cooled by forcer— Cooling method in which the bulbs are cooled in their packing cases by the forcer.

Case cooled by supplier—Cooling method in which the bulbs are cooled in their packing cases by the supplier before they are shipped to the customer. The forcer pots the bulbs upon arrival and begins the forcing process.

Cation exchange capacity—The measure of a media's capacity to hold nutrients.

Catkin—A spike, raceme, or cyme composed of unisexual flowers without petals and falling as a unit.

Cell packs—Molded plastic containers divided into two, three, four, or six separate growing compartments used for bedding plants.

Cellular respiration—The process in which sugars made in photosynthesis are broken down into simpler molecules. The energy is applied towards growth and development of the plant.

Cellulose—Huge molecules used in cell walls for strength and rigidity that result from the bonding of thousands of glucose molecules. Once sugars are converted to cellulose, they are not reclaimed for other purposes.

Center bud removal—Method of disbudding which involves the pinching off of the most terminal flower bud.

Chemical control—The use of chemicals to control pests and diseases.

Chlorophyll—A green pigment that absorbs light and is a key component of the photosynthetic process.

Chloroplasts—Specialized organelles within the individual plant cells that harness light energy. They contain chlorophyll, carotene, and xanthophyll pigments.

Chrysanthemum—Potted flowering plant originating in China, Japan, and Europe. These short-day plants come in many colors and the flower forms differ.

Cineraria—A cool season potted flowering crop with daisy form flowers held in a cluster above the foliage. They come in blues, pinks, reds, and white.

Clone—Offspring that are the genetic duplicates of the parent plant.

Coir—Growing media component that is a waste product of the coconut industry and a renewable resource. Its characteristics are similar to those of peat moss.

Complete flowers—Flowers that have sepals, petals, stamens, and a pistil.

Compound leaves—A petiole and two or more leaf blades called leaflets.

Computer controls—Environmental controls that utilize microprocessors. The microprocessors make complex judgements based on information from a number of sensors. Those sensors might gather information on temperature, relative humidity, and sunlight.

Computerized environmental management—Expensive environmental control systems that are accurate and offer the greatest range of uses. All pieces of automated equipment can be controlled together. For instance, vents may be open or closed at the same time curtains are open or closed. Computerized environmental management systems also can be programmed to provide different zones in the greenhouse different climate conditions.

Conditioning—Preparation of plant materials before arranging them by allowing for uptake of water or preservative solution to extend the life of the flowers.

Conservatory—A greenhouse designed for the display of plants.

Controlled temperature forcing (CTF)—Method of cooling whereby bulbs are received by the forcer and potted immediately.

Cooperation—Working with everyone for the common purpose.

Cormels—Miniature corms that can be separated and planted.

Corm—A short, swollen, underground stem. Crocus and gladiolas are corms.

Corolla—The collection of petals on a flower.

Corymb—A more or less flat-topped inflorescence having a main vertical axis and pedicels or branches of unequal length.

Corporation— Business structure with a legal association of individuals.

Cotyledon—Seed leaves. Embryos of monocot plants have one cotyledon, while those of dicot plants have two cotyledons. Monocot plants store the bulk of their energy in the endosperm. Dicots store their food in the two cotyledons.

Crocus—Bulb crop that is actually a corm and is native to Central and Western Europe. Crocus varieties offer flowers of bright purple, yellow, and white colors.

Cross-pollination—When the pollen of a plant pollinates the flower of another plant.

Cultural/physical control—Methods that physically prevent activities of pests.

Cultural practices—The control or management of light, temperature, air, and water to promote healthy growth.

Cut flowers—Flowers grown in a greenhouse or field, cut when they reach a certain maturity, and sold to a wholesaler. Roses, carnations, chrysanthemums, and orchids are a few common cut flowers.

Cut cultivated greens —Leaves harvested for floral design work, often referred to as "greens." Leatherleaf, lemonleaf pittosporum, podocarpus, asparagus fern, and huckleberry are common foliage materials used in floral work.

Cuticle—The waxy coating on epidermis cells that serve to prevent excessive water loss from the leaf tissues.

Cuttings—One of the most common and simplest methods of asexual propagation in which portions of stems, leaves, or roots are used.

Cyclamen—Potted flowering plant native to the Mediterranean. The unusual shaped flowers come in white, pink, carmine, and red. Cyclamen are considered a cool crop.

Cyme—Usually a flat-topped inflorescence.

Cytokinins—Hormones responsible for cell division and differentiation. They are produced in the roots and play a role in apical dominance of intact plants. While auxins inhibit the growth of lateral shoots, cytokinins promote lateral shoots. In roots the situation is reversed. Auxins promote the branching of roots, and cytokinins inhibit branching.

Daffodil—Also known as narcissus. This bulb produces brilliant yellow flowers, with some varieties available in white, two-tone colors, and peach.

Damping off—A term used to describe the early death of seedlings that have germinated. A number of fungi can be the cause.

Day-neutral plants—Plants unaffected by day length.

Dependability—Term that means worthy of trust and that the employee can be counted on to do a job as instructed.

Determinate—Term used when the first flower of a inflorescence to open is at the apex of the stem and the progression of flowering is downward or outward.

Dicot—One of two main types of plants. Dicotyledon plants or dicot plants have xylem and phloem in a ring within the stem. Examples of dicots are hydrangea, florist azalea, African violet, and chrysanthemum.

DIF—The mathematical difference between the day temperature and the night temperature.

Diploid—The nuclei of normal cells that contain a double set of chromosomes.

Disbudding—The removal of flower buds to improve the overall quality of the plants.

Disease—A disturbance to the normal, healthy growth and development of a plant.

Dish garden—Differing types of potted plants in the same container.

Division—Asexual propagation in which the plant roots or the entire plant may be cut into sections to make two or more plants from the original plant.

Double fertilization—Process in flowering plants whereby both sperm nuclei in the pollen grain are involved in fertilization.

Drench—Pesticide application in which material, mixed with water, is simply poured into the pots.

Drip irrigation—Drip irrigation is similar to spaghetti tubing except the tube provides a slow steady drip with little run-off.

Dwarf iris—Dwarf iris is a bulb crop native to the Caucasus region. They are purple in color.

Easter lily—Bulbs grown in the United States and Canada as a potted flowering crop for the

Ebb and flood—Watering method in which water is pumped into the system at regular intervals, filling the bench. The pots sit in the water while the media in the pots slowly absorbs water. After a certain period, the water drains from the bench.

Electromagnetic spectrum—Radiant energy given off by the Sun has wavelengths that are measured in nanometers, and based on the measurements have been placed on a scale or spectrum.

Embryo—An immature plant.

Endosperm—Tissue in which food is stored for the embryo.

Energy curtains—Automated systems utilizing fabrics that insulate the greenhouse at night and shade the crops during the day.

Enthusiasm—Term used to describe workers as being lively, excited, involved, and interested in their work.

Enzymes—Chemical activators with specific jobs. With split-second timing they break down sugars and recombine them with nitrogen and other minerals to produce many complex molecules are produced.

Epidermis—The protective layer of cells on leaves.

Erwinia—A bacterial disease that causes rotting of plant tissues. The bacteria enter the plant through wounds. It is a common problem on *Dieffenbachia*, *Philodendron*, chrysanthemum, and cyclamen

Ethylene—A colorless gas produced in the nodes of stems, ripening fruits, and dying leaves. It causes the thickening of stems and speeds the aging of plant parts, particularly fruit.

Ethylene inhibitors—Treatments or chemicals that block or tie up the ethylene within the flower and reduce its impact.

Etiolation—Term used to describe a severe case of stretching of plants grown in low light or darkness.

Even-span greenhouse—Greenhouse structure that has a roof with an even pitch and an even width. It is a single house.

Explant—Small pieces of plant material.

Fan and pad cooling system—Cooling system based on the evaporation of water. This system involves a wall of cellulose or aspen pads on one end of the greenhouse. The pads are kept wet by a system of pumps and gutters that re-circulate the water. On the other end of the greenhouse there are a series of fans. The fans pull air through the pads and across the greenhouse. As the air passes through the pads, the water evaporates.

Fan-jet tubes—Polyethylene tubes that run the length of the greenhouse, and they have holes along the side of the tube. The fan-jet tubes are attached to fans associated with unit heaters. When the fans are operating, the tubes inflate. Air is blown through the holes in the tube. Fan-jet tubes promote an even circulation of air in the greenhouse.

Far-red light—A wavelength that plays an important role in plant growth and development as it triggers a shade avoidance response in plants when levels of blue or red light are low.

Fertigation—The practice of fertilizing while irrigating.

Fertilization—The fusion of a sperm with an egg.

Fertilizer—Any material added to growing media that provides nutrients for plants.

Fertilizer injector system—Standard equipment in today's greenhouses. A concentrate of water-soluble fertilizer is mixed in a tank in exact proportions before irrigating plants. With fertilizer injectors growers can provide exact levels of nutrients.

Fiberglass—Glazing material widely used in the 1960s and 1970s. Light transmission on new sheets is good, but it discolors after 7 to 10 years. Hail can cause some damage. Fiberglass is highly flammable.

Fictitious name statement—A document that must be filed with the county to register a business name and prevent another business from using that name.

Filament—The stalk of the stamen that holds the anther.

Finished—Grown to saleable size.

Flavonol—A pigment that produces yellow and cream colors.

Floral arranger—People with the training required to make and copy floral designs.

Floral design—The art of organizing the design elements inherent in plant materials, container, and accessories according to the principles of design to attain a composition with the objectives of beauty, simplicity, harmony, suitability, and expression.

Floral designer—Person considered to be more skilled than an arranger in that he or she has the ability to create and design unique floral art.

Floral foam—A porous material that holds water and provides stability for flower stems.

Floral foam holder—A wedding mechanic with floral foam encased in a plastic cage with a handle.

Floral production—The growing of flowering or foliage crops to sale size or maturity.

Floriculture—Literally defined as the "culture of flowers," floriculture is an international, multi-billion dollar industry based on flowering and foliage (leafy) plants.

Florist—Person who works with cut flowers and foliage.

Florist azalea—Two species of *Rhododendron* are grown as florist azaleas. Different varieties are grown for Christmas, Easter, and Mother's Day.

Florist hydrangea—Florist hydrangeas are native to Japan. They are grown primarily for Easter and Mother's Day. The color of their long-lasting flowers (blue, pink, white or red) is influenced by pH.

Flowers—The reproductive organs of plants.

Focal area—A center of interest.

Focal point—A center of interest.

Fog system—Cooling system that involves an atomizer that produces water vapor. The water vapor cools the house by flash evaporation. Fog systems increase the humidity within the house while they cool.

Foliage plants—Tropical and subtropical plants grown for their leaves rather than for their flowers and selected for their ability to be grown indoors. They are also called houseplants.

Foot candle—A measure of light intensity based on the amount of light distributed by a single candle one foot away.

Forcing—A term used to describe practices that get bulbs to grow and produce flowers.

Freesia—A cut flower with very fragrant lily-like flowers, borne on a horizontal spike. They come in bright colors including yellow, red, white, blue, and pink.

Fumigant—Poisonous gases that are distributed through the air to all parts of the greenhouse.

Fungicide—Pesticide used to control fungus.

Fungus gnats—Long-legged, winged, gray-black insects less than an eighth of an inch long. The larva of fungus gnats feed on root hairs and tunnel into plant stems.

Genetic engineering—Selecting and moving genetic material from one plant to another.

Germination—The beginning of growth from a seed.

Germination rate—The percentage of seeds capable of germinating.

Gibberellins—Hormones that induce stem cell elongation and cell division. Gibberellins play a key role in stimulating the development of flowers. They also have an important role in the production of enzymes during seed germination.

Gladiolus—Cut flower valued for their large showy floral spikes. Cultivars are available in nearly every color. As cut flowers, they are long lasting.

Glass—Considered to be the best glazing material for plant production. It provides the highest light transmission of any glazing material at 88% to 89%. The initial cost of glass is expensive, but it is very long lasting, provided it doesn't break.

Glazing—The covering of a greenhouse.

Gloxinia—Potted flowering plant native to Brazil with large, trumpet-shaped flowers that come in red, pink, purple, white, and two-tones. Gloxinias are propagated from their tubers or by seed.

Grades—Flowers grouped according to size or quality.

Grafting—The process in which the stem of one plant is made to grow on the roots of another plant.

Granular—Pelleted form of pesticide that is carefully measured and applied on the surface of the growing medium.

Grape-hyacinth—Bulb crop native to Asia Minor. Grape-hyacinths have clusters of blue or white flowers. The blues can sometimes appear iridescent.

Gravitropism—A plant's response to the force of gravity.

Greenhouse—A structure enclosed by glass or plastic that allows light transmission for plant growth.

Greenhouse sanitation—The efforts made to keep a greenhouse clean.

Grower—The person in charge of the crop production.

Growing media—The material in which plants are grown.

Growth retardants—Chemicals that inhibit the action of gibberellins on stem elongation.

Guard cells—The stomata are opened and closed by a pair of guard cells. When water is plentiful and light is shining, the guard cells are pumped full of water. This causes the cells to push apart, creating an opening to allow an exchange of gases.

Haploid—Reproductive cells, sperm and egg, that have a single set of chromosomes.

Hard baskets—Hanging baskets that contain plants that are propagated by cuttings, such as fuchsia, geraniums, and New Guinea impatiens.

Hard pinch—The removal of everything but 3 or 4 leaves by pinching.

Hard water—Water that contains high amounts of minerals.

Head—A rounded or flat-topped cluster of sessile flowers.

Headhouse—A structure attached to the greenhouse. It serves as a storage area for growing media, containers, fertilizers, and other materials. It is a work area.

Heat delay—A delay of flowering by several days to weeks because of high temperatures.

Herbicide—Pesticide used to control weeds.

High-intensity discharge (hid) lighting system—New efficient and effective light sources.

Highly buffered—Resistant to a change in pH.

High-visibility colors—Warm hues, light values (tints), and bright intensities which are very noticeable from a distance.

Holiday cactus—Potted flowering plant with three types available—Thanksgiving, Christmas, and Easter. The natural photoperiodic response of these plants brings them to flower around the holiday for which they are named.

Honesty—reflects truthfulness, integrity and fairness.

Hook cut—The removal of a portion of the old wood with a flower when roses are harvested.

Hoop-house—Even-span greenhouse that has an arching pipe framework.

Hormones—Natural chemicals produced by plants to regulate growth.

Horticultural oils—Material sprayed on the insects to clog the breathing pores of the insect, causing suffocation.

Hot water system—Heating system that uses water heated in boilers and pumped through pipes in the greenhouse. The pipes are most often placed under the benches or in the floor. Hot water systems have relatively low maintenance.

Hyacinth—Bulb crop native to Greece and Asia Minor. They produce a fragrant flower that may be purple, white, pink, or multi-colored. Most hyacinths are produced in Holland.

Hybrid—The offspring of two plants of the same or related species that differ genetically.

Hybridization—The process in which plant breeders select plants for outstanding characteristics, collect pollen from one plant and transfer it to the flower of another in the hope of having the outstanding characteristics expressed in the offspring.

Hybrid tea rose—A large, usually solitary flower. The majority of the hybrid tea roses produced are red. Other popular colors include pink, yellow, and white.

Image—Term used to describe the impression people have of a business.

Imperfect flowers—Flowers that lack one of the two sex structures.

Incomplete flowers—Flowers that lacks any one of the following; sepals, petals, stamens, and pistil.

Indeterminate—Term used when the last flower of an inflorescence to open is terminal on the main axis and the progression of flowering in inward or upward.

Indoleacetic acid (IAA)—The primary auxin.

Indolebutyric acid (IBA)—Widely used synthetic root-promoting material.

Infectious disease—A disease that can be spread to other plants.

Inflorescence—Flower clusters, formed by the branching system of the stem.

Infrared heat system—Heating system in which infrared wavelengths are absorbed by the plants, media, and benches. The heat then transfers to the air space around them. Therefore, infrared heat does not heat the air directly.

Initiative—Moving on to a task without being instructed to do so.

Insect growth regulators—Materials that disrupt the growth and development of the insects.

Insecticidal soaps—Materials that dissolve the protective membranes of insects, bringing on death.

Insecticide—Pesticide used to control insects.

Integrated pest management (IPM)—Use of a wide range of strategies to control pests and diseases.

Interior plantscaping—The use of plant materials to improve the appearance of the indoor environment.

Intermittent mist system—Watering method used to deliver water in tiny droplets for the purpose of keeping plant material moist. Intermittent mist systems are used for plant propagation.

Irrigation—Furnishing plants with water by flooding or sprinkling.

Irrigation booms—Watering method involving booms with nozzles that sweep across greenhouse benches delivering overhead water.

Kalanchoe—Succulent short-day plants native to Africa and Asia. Their flowers range in color from bright yellow to orange to hot pink.

Lateral bud—Buds located along the sides of a stem.

Layering—A method of asexual reproduction whereby roots form on a stem while the stem is still attached to the parent plant.

Leaching—The process in which soluble salts are washed from a growing medium as water flows through the growing media.

Leadership—The ability to influence other people to meet individual or group goals.

Leaf blade—The large broad part of a leaf.

Leaf-bud cutting—Cuttings composed of the leaf blade, the petiole, a bud at the base of the petiole, and a portion of the stem.

Leaf counting—Practice used to monitor the progress of an Easter lily crop and to determine whether growth needs to be speeded or slowed.

Leaf cutting—Entire leaves or portions of a leaf are removed from the parent plant for use as cuttings.

Leaf margin—The edge of the leaf blade. The margins can be one of many forms including wavy, toothed, lobed, and entire or smooth.

Leaf miners—Small stocky flies whose larva feed on the interior of the leaf, making tunnels as it moves along.

Leaf-petiole cutting—The leaf blade and the petiole are taken from the parent plant for use as a cutting.

Lean-to greenhouse—A greenhouse structure that is attached to a building, with the building providing support for the roof.

Leaves—Plant organs responsible for the production of food (sugars) for the plant.

Liners—Small-branched plants are sold as 4", 6", or 8" diameter in azalea production.

Long-day plants—Plants that flower as days lengthen in the summer.

Loyalty—Faithfulness to the company.

Macronutrient—Six of the 17 essential elements that are required in greater quantity than the others including nitrogen (N), phosphorus (P), potassium (K), calcium (Ca), magnesium (Mg), and sulfur (S).

Marketing—The total activity associated with the advertising, promoting, selling, and shipping of products and services.

Mealybugs—Slow moving, oval-shaped, whitish insects that have a waxy finish and produce small cottony masses.

Mesophyll—Photosynthetic tissue sandwiched between the epidermal. The bulk of photosynthetic activity for a plant takes place in the mesophyll cells.

Metabolism—Term used to describe all of the chemical reactions in a plant.

Micronutrient—Sometimes they are called trace elements because they are needed in small amounts. They are boron (B), copper (Cu), chlorine (Cl), iron (Fe), manganese (Mn), molybdenum (Mo), sodium (Na), nickel (Ni), and zinc (Zn).

Mineral soil—Growing media component obtained from nature and is the result of weathered rock. It has become more difficult for growers to find uniform mineral soil with a favorable pH and good structure and texture.

Mini carnation—One of three types of carnations. Mini carnations are ½ to $^1/_3$ the size of standard carnation flowers.

Mites—Tiny eight legged bugs related to spiders. Mites pierce plant leaf tissues and suck juice.

Miticide—Pesticide used to control mites.

Monocot—One of two ma n types of plants. Monocotyledons or monocot plants have xylem and phloem scattered throughout the stem and tend to have narrow leaves. Easter lilies, iris, spider plants, and dracaenas are examples of monocotyledons.

Motivated—employees who take action and have a desire to do a good job and to improve themselves.

Mottled foliage—Leaves with flecks of a differing color throughout the entire leaf.

Naphthaleneacetic acid (NAA)—Widely used synthetic root-promoting material.

Narcissus—See daffodil.

National FFA Organization—The premier student organization for young people interested in the greenhouse industry.

National Junior Horticultural Association (NJHA)—An organization for young people interested in the greenhouse industry.

Natural cooling—Method of cooling whereby bulbs are potted as soon as they arrive. They are placed outdoors or may be kept in a poly house if temperatures drop below freezing.

Non-infectious disease—Diseases caused by environmental imbalances and cannot be spread to other plants.

Non-precooled—Bulbs that are not given cold treatment by the supplier.

Non-tunicate bulbs—Bulbs that lack a tunic and the protection a dry, papery covering would provide.

Nutrient deficiency—If a plant fails to receive the needed amount of nutrients, it will show signs of an unhealthy plant appearance. Symptoms vary with the nutrient that is in short supply.

Organic matter—Decayed or partially decayed remains of plants and animals found in the media.

Oriental lilies—Bulb crop native to the Asian Continent. They come in a wide range of colors, including yellow, white, red, and orange. They are forced throughout the year.

Ornamental horticulture—The practice of growing and using plants for decorative purposes.

Ovary—The part of the pistil that contains one or more ovules.

Overhead sprinklers—Sprinkler nozzles in a fixed position above the crops deliver overhead water.

Ovule—The part of the flower in which the eggs are produced and seeds develop.

Palisade layer—A stacked layer of mesophyll cells found just below the upper epidermis.

Panicle—A more or less elongated inflorescence with a central axis along which there are branches that are themselves branched.

Partnership—Business structure consisting of a number of individuals who have agreed to share the responsibilities of the business.

Parts per million—A measure of nutrients in solution.

Pathogens—Disease-causing organisms such as bacteria, fungi, or viruses.

Peat moss—Widely used organic growing media component dug from peat bogs. Peat moss has light bulk density, good moisture holding ability, good air space qualities for the exchange of gases, adequate cation exchange capacity, and a stable pH that is usually between 3.5 and 4.5.

Pedicel—The stem of a flower.

Peduncle—The main stem of an inflorescence.

Perfect flowers—Flowers that have both the male and female parts.

Perianth—Term used to describe the sepals and petals together.

Perlite—Lightweight, white growing media component originating from volcanic rock. Perlite is stable, sterile, has little cation exchange capacity, has a pH of 7.5, and provides good water drainage and aeration. Perlite contains fluorides that cause leaf damage to some monocotyledon plants.

Persian violet—Persian violet is from the Island of Socotra off the tip of Somalia. It has pretty violet-blue flowers that cover the whole plant.

Personal growth—Deals with social skills, citizenship, and topics associated with self-esteem. Pest—A living organism that can cause injury or loss to a plant.

Pesticide—Chemical used to control pests and diseases.

Petals—Structures located just inside the sepals. Petals appear leaf-like and are often very colorful to attract pollinators.

Petiole—The leaf stalk that connects the leaf blade to the stem.

Phloem—Conductive tissue of the stem in which food made in the leaves is transported to the rest of the plant.

Photoperiodism—A plant's response to light duration.

Photosynthesis—The process of a plant making sugars (food) from water and carbon dioxide. Photosynthesis takes place in the leaves. Growth and development of greenhouse crops is determined by the rate of photosynthesis.

Phototropism—The ability of a plant to bend towards a light source.

Phytophthora—A fungal disease similar to *Pythium*. It causes crown and stem rots in cool, wet conditions.

Pinching—Removal of the apical meristem. Pinching is a common practice used to produce bushy, well-branched crops.

Pistil—The female part of the flower.

Plant growth regulators (PGR)—Natural occurring or synthetic chemicals that regulate all growth and development of plants.

Plant health—Refers to plants that are free of pests and disease.

Plantlet—Shoots that develop from buds in tissue culture.

Plant nutrition—The absorption of nutrients available in the growing media for plant growth. Current fertilizer practices allow growers to adjust and deliver nutrients to maintain desired plant growth.

Plant propagation—The reproduction of plants by seed or by a portion of a plant.

Plugs—Small plants grown in a small amount of media in divided trays.

Pocketbook plant—Native from Mexico to Chile, the flowers this cool-season potted flowering crop resemble a woman's purse. The flowers are very colorful yellows, oranges, and reds.

Poinsettia—Plant native to Mexico associated with Christmas and sold as a potted plant only during that time of the year.

Pollen—Produced by the anther, contains the male sex cells.

Pollination—The transfer of pollen to the female part of a flower.

Polycarbonate structured sheets—The most widely used structured sheet, commonly manufactured with a "twin wall" held together by ribs. Polycarbonates have a good 80% light transmission. They resist hail damage.

Polyethylene—Flexible plastic sheets, usually 6-mil in thickness, used as a glazing material by stretching it over the greenhouse framework.

Polyploid—Plants that have more than the normal double set of chromosomes.

Porosity—The percentage of pores or spaces between the solid particles of a growing media.

Pot fillers—Automated greenhouse systems that helps workers fill pots or flats with growing media.

Potted flowering plants—Plants grown in pots for their showy flowers. Some popular potted flowering plants are poinsettias, chrysanthemums, cyclamen, and African violets.

Powdery mildew—A fungal disease. Characteristic symptoms include a white dusty coating to leaves and flowers. Infected plants can become stunted.

Precooled—The process of providing cold treatment to bulbs for early forcing, or quickly cooling flowers after harvesting and packaging and prior to shipping.

Primary root—A single, main root.

Primrose—Six species of this cool-season potted flowering crop are used in greenhouse production. All of these originated in the northern temperate regions of the world. Primroses are available in a wide range of bright, colorful, and fragrant flowers.

Pruning shears—A cutting tool with sharp cutting blades that will cut thick or woody plant material without pinching.

Pythium—A fungal disease that attacks greenhouse plants under cool, wet conditions.

Raceme—An elongated inflorescence with a central axis along which are simple pedicels of more or less equal length.

Receptacle—The tip of the stem that holds the flower parts.

Respiration—The process of the plant using its stored food.

Retail florist—Provides floral design services, cut flowers and other floriculture crops. Retail florists have typically been small independent businesses.

Retractable-roof greenhouse—Greenhouse structures with roofs that can be opened and closed. They allow the grower to open the house to the elements when weather is favorable for plant growth and close the roof when the crops need protection.

Rhizoctonia—A fungal disease prevalent under wet and warm conditions.

Rhizome—An underground horizontal stem. Iris and calla lilies are rhizomes.

Ridge-and-furrow greenhouse—Greenhouse structures that consist of a number of greenhouses connected together along the length of the houses. Shared side walls are eliminated, creating a large interior space.

Rock wool—A human-made growing media component made from an igneous rock, basalt. It consists of fibers similar in appearance to cotton candy. Rock wool is used extensively in hydroponic operations. Rock wool has good water holding capacity and good aeration. It is slightly alkaline and has a low cation exchange capacity.

Rolling bench—Benches rest on pipes allowing each bench to be rolled from side to side. With this system only enough space needs to be kept for one aisle.

Roll out pinch—The removal of just the meristematic tip of the stem.

Root cap—A mass of cells on the root tip that protects the root from coarse soil.

Root hairs—Tiny hairs found near the growing tip of the root. Root hairs greatly increase the surface area of the root allowing more water and minerals to be absorbed into the plant.

Rooting room A—Involves bulb production for early flowering, before February 14.

Rooting room B—Involves bulb production for later flowering, after February 14.

Rootstock—The lower portion of the graft that includes the root system.

Root system—Term used when the roots are discussed collectively.

Root tuber—An enlarged food-storage root with adventitious shoots.

Sales—The exchange of a product for money.

Salesperson—Person who has direct and usually frequent contact with the retail florist. The salesperson takes orders and arranges delivery of the products.

Salinity—Salt content.

Sand—Natural growing media component that is a result of the wearing of rock. Sand is heavy and has a high bulk density that provides solid support for larger plants. It improves water drainage and aeration when used with soil. The pH of sand is between 7.5 and 8.5.

Saucers—Saucers are used with overhead irrigation systems in a manner similar to capillary mats. As water is applied, it is collected in the saucer and drawn up into the growing media by capillary action.

Scale—Insects with flat, oval, often brown bodies. They may or may not be covered with an armored shell. Scale insects pierce plant leaves and stems and suck juices. Also, the relative size among objects or parts of objects.

Scales—Fleshy modified leaves of a bulb.

Scalettes—Small bulbs that grow on bulb scales that were removed from a bulb and planted.

Scarification—Dormancy mechanism that requires the breaking down of the seed coat for germination to occur.

Scion—The portion of the graft that is to become the stem.

Scooping—The removal of the basal plate of the bulb and the base of all of the bulb scales.

Scoring—Similar to scooping, however, the basal plate is not removed. Two cuts that cross the basal plate are made about a 1/4 inch deep.

Secondary root—A smaller root that branches off the primary root.

Seed—Structures that have an embryo and a source of stored food contained within a seed coat.

Seed coat—A protective shell that protects the embryo and the endosperm from drying and from physical injury.

Self-pollination—When the pollen of a plant pollinates a flower on the same plant.

Senescence—The aging of plants.

Sepals—Green, leaf-like structures on the exterior of the flower. The sepals fold back as the flower opens.

Separation—Vegetative plant structures that are removed intact from the parent plant.

Sexual propagation—Reproduction that involves flowers and seeds. In sexual propagation, flowers are pollinated and eggs fertilized. The resulting offspring is genetically different from the parent plants.

Shattering—The breaking apart or the falling out of petals.

Shelf life—Refers to the time the plant maintains its health while on display for sale.

Short-day curtains—Curtains that block light from reaching the crops. When closed for 12 to 13 hours a night, they produce a short-day (actually a long-night) effect.

Short-day plants—Plants that will begin to flower naturally in the fall when the day lengths get shorter.

Silica gel—A type of drying agent available as white crystals or color-indicator crystals that change colors when the desiccant is too moist to effectively use.

Silica sand—A type of drying agent, white builder's sand.

Simple leaves—A single leaf blade and a petiole.

Single flowering carnation—One of three types of carnations. Produce a flower without the need of disbudding.

Slow-release fertilizers—Fertilizers that continually discharge a small amount of nutrients over a period of time.

Snapdragon—Cut flower with an elongated floral inflorescence that comes in a variety of bright colors. White, pink, and yellow are very popular. Other colors available include red, bronze, and purple.

Soft baskets—Hanging baskets that contain plants that are propagated from seed.

Softened water—Water that has been treated with salts to remove minerals.

Soft pinch—The removal of ½" to ¾" of the stem.

Soil-less mix—Growing media that lack any mineral soil.

Soil pH—The measure of acidity or alkalinity of the media.

Sole proprietorship—Business structure in which the assets and liabilities belong to the owner or owners. The owner makes decisions and receives all profits.

Solitary flower—Flowers borne one to a stem. Examples include tulip, narcissus, and rose.

Soluble salts—Dissolved mineral salts that can be harmful to plants. Nutrients in solution along with mineral salts from the water and media contribute to the soluble salt content in the growing media.

Sow—Planting of seed.

Spadix—A spike with a thickened, fleshy axis, usually enveloped by a showy bract called a spathe.

Spaghetti tubing—A watering method involving small tubes connected to a main line. The end of each tube is placed in an individual pot. When operating, water dribbles through the tubes, watering all the pots on a bench simultaneously.

Spike—An elongated inflorescence with a central axis along which are sessile flowers.

Splitting—A condition associated with hyacinths where the flower stalk separates from the basal plate.

Splitting of the calyx—The biggest problem with harvested carnations. This condition, when the calyx splits, petals spill out and the flower appears misshapen, is caused by fluctuations of temperatures and is more common with some cultivars than others.

Spongy layer—A loosely arranged layer of mesophyll cells underneath the palisade layer.

Spray—Pesticides applied by small droplets of the spray that cover the surface of the plant.

Stamens—Male reproductive parts of a flower.

Standard carnation—One of three types of carnations. A large, solitary flower that has been disbudded.

Starch—Huge molecules in which food is stored for plants resulting from the bonding of thousands of glucose molecules.

Steam heat—Heating system involving large boilers that bring water to a boil to produce steam. The steam then flows through pipes in the greenhouse. Steam heat is not as uniform as hot water heat. Nor is it as easy to adjust the temperature.

Stem cuttings—Method of cutting in which stems are taken 2 to 5 inches in length.

Stem tuber—Swollen tips of a rhizome. The Irish potato is a stem tuber.

Sticking cuttings—The practice of placing cuttings in rooting cubes or directly into a media.

Stigma—Found at the end of the pistil and it has a sticky surface on which pollen can be caught.

Stock plant—Plants from which cuttings are removed.

Stolon—A stem that grows horizontally above the ground and may produce roots at its tip or at nodes. Airplane or spider plants and strawberry begonia are two foliage plants that produce stolons.

Stomata—The pores or openings in the epidermis that allow the exchange of oxygen, carbon dioxide, and water vapor.

Stratification—The process whereby a seed must go through a period of cold temperatures before it will germinate.

Style—The neck of the pistil.

Styrofoam—Growing media component made of synthetic polystyrene material. It is very lightweight and stable. While it provides good drainage and aeration, it lacks cation exchange capacity, water holding abilities, and has a neutral pH.

Supervised experience programs (SE)—Valuable real-life work experiences obtained outside the classroom.

Sweetheart (floribunda) rose—A smaller rose that grows as a spray and is often disbudded. Sweetheart roses are used extensively in wedding work, corsages, and centerpieces. Yellow and pink make up a large percentage of those produced. New cultivars also offer many pastel colors.

Synthetic growth regulators—Chemical compounds that regulate growth but are not produced by plants.

Tax number—Another term for resale license or permit.

Tepal—Term used when a clear distinction between petals and sepals cannot be made.

Terminal bud—The large bud at the tip of a stem.

Tetraploid—Plants with four sets of chromosomes.

Thermoperiodism—Term used to describe a temperature requirement that produces a plant response.

Thermostats—Low cost and easy to install environmental controls. There are two types, on-off and proportioning thermostats. On-off thermostats control fans, heaters, vents with a change of temperature. Proportioning thermostats provide continuous control of systems with changes in temperature.

Thielaviopsis—A fungus that causes root and stem rots.

Tissue culture—A very technical method of asexual propagation that involves the culture or growing of small pieces of plant tissue.

Toning—A term used to describe procedures used to prepare plants for a post-production environment.

Transpiration—The movement of water vapor through stomata.

Transplanted—The process of moving plants from one container to another or from a container to the ground.

Tray mechanization—Also known as Dutch trays or palletized benches, this system consists of benches that hold 100 or more pots. The trays are rolled on a series of rails from one area of the greenhouse operation to another.

Triploid—Plants that have three sets of chromosomes.

Tulip—Bulb plant from southwestern Europe and the Near East. They are available in practically all colors with the exception of a true blue, and are grown for cut flowers or as potted flowering plants.

Tunic—A dry, papery covering.

Tunicate bulb—A bulb with a dry, papery covering.

Turgid—Tissues full of water.

Turgidity—Water pressure in the plant cells.

Umbel—An inflorescence having several branches arising from a common point.

Unavailable water—The thin film of water that binds so tightly on media particles that it can not be used by plants.

Uneven-span greenhouse—A greenhouse structure that has a roof with an unequal pitch and width. Use of the uneven span structure is limited to hillsides.

Unit heaters—Heating system with which air is heated within a unit and then blown by fans throughout greenhouse. Unit heaters are less expensive to install than hot water systems.

Variegated foliage—Leaves with lines, stripes, or areas of a differing color along the leaf edge or center.

Variegation—The different color patterns of leaves caused by pigments.

Variety—A diverse mixture or differing components.

Vase life—The length of time cut flowers and foliage live after they have been cut.

Vents—Panels that open and allow air exchange with the outside.

Vermiculite—Common growing media component made from the mineral, mica, that has a good cation exchange capacity and a neutral to slightly alkaline pH (6.3–7.8).

Vernalization—A physiological process whereby a period of cold temperature is required for flowering.

Viability—The ability of seeds to germinate under optimum conditions.

Vigor—The ability of seeds to germinate under different conditions and still produce healthy seedlings.

Virus—Symptoms of infection include discoloration of plant tissues, stunting of growth, and deformed growth. Tobacco mosaic and Aster yellows are two common viral disease associated with greenhouse crops.

Water-soluble fertilizers—Fertilizers that dissolve completely in water and stay in solution.

Water tubes—Small, rubber-capped, plastic tubes which hold water and a single flower or cluster of small flowers.

Western flower thrips—Small insects with two pairs of fringed wings. They are dark brown in color.

Whitefly—Small white insects that camp out on the undersides of leaves where they pierce the tissues and suck juices. Their flat, scale-like larva feed on the undersides of the leaves.

Wholesale floral manager—Person who oversees the operations of the business by supervising the staff, estimating retail demands for products, and preparing long-range plans.

Wholesale florist—A business that buys from growers or brokers and sells to retail florists at wholesale prices.

Wilting—A drooping condition and a lack of firmness to the plant tissues caused by inadequate water.

Worms—Larva of various moth species. They damage greenhouse crops by eating the plants.

Xanthophyll—A pigment with colors that range from yellow to nearly no color at all.

Xylem—The conductive tissue in the stem that transports water and minerals from the roots to the leaves.

Yellow sticky traps—Brightly colored traps on which insects land on, get stuck, and die. They are effective in monitoring pest populations.

Zygote—A fertilized egg.

Bibliography

Armitage, Allan M., *Bedding Plants: Prolonging Shelf Performance.* Batavia, IL: Ball Publishing, 1993.

Ball Redbook, 16th ed., edited by Vic Ball. Batavia, IL: Ball Publishing, 1998.

Capon, Brian. *Botany for Gardeners*. Portland, OR: Timber Press, 1990.

Floriculture and Environmental Horticulture Situation and Outlook Report. United States Department of Agriculture, Economic Research Service, October, 2001.

Floriculture Crops: 2001 Summary. United States Department of Agriculture, National Agriculture Statistics Service, 2002.

Greenhouse Business. Park Ridge, IL: McCormick Communications Group.

Greenhouse Product News. Des Plaines, IL: Scranton Gillette Communications.

Grey, Don. "Equipping Your Retractable-Roof Greenhouse," *Grower Talks Magazine*, January, 1997.

Grower Talks on Plugs II, 2nd ed., edited by Debbie Hamrick. Batavia, IL: Ball Publishing, 1996.

Hartmann, Hudson T., Dale Kester, Fred Davies, and Robert Geneve. *Hartmann and Kester's Plant Propagation: Principles and Practices,* 7th ed. Upper Saddle River, NJ: Prentice Hall, 2002.

Kimbrell, Grady, and Ben S. Vineyard. *Succeeding in the World of Work.* Mission Hills, CA: Glencoe Publishing Company, 1981.

Lectures by and discussions with Dr. Marvin Carbonneau, Professor Emeritus, Ornamental Horticulture, University of Illinois.

A *Manual for Flower Judging.* Prepared by Pi Alpha Xi National Floriculture and Ornamental Horticulture Honor Society; edited by the Department of Horticulture, University of Maryland; printed by Department of Ornamental Horticulture, California Polytechnic State University, San Luis Obispo, CA.

Pilon, Paul. "Adding a Perennial Crop: The Logistics and the How-Tos," *Greenhouse Product News,* November, 2001.

Ranch, Paul Ecke. *Technical Information Bulletins.* Encinitas, CA: 2002.

Schroeder, Charles B., Eddie Dean Seagle, Lorie M. Felton, John M. Ruter, William Terry Kelley, and Gerard Krewer. *Introduction to Horticulture,* 4th ed. Upper Saddle River, NJ: Prentice Hall Interstate, 2004.

Solomon, Eldra, Linda Berg, and Diana W. Martin. *Biology,* 6th ed. Brooks/Cole Publishing, 2002.

Svenson, Sven E. "Advantages of Open-Roof Greenhouses." Oregon State University.

Walt, Vivienne. "Flower Trade," *National Geographic Magazine*, April, 2001.

Water, Media, and Nutrition for Greenhouse Crops, edited by David Wm. Reed. Batavia, IL: Ball Publishing, 1996.

Index